MARRIED to TELEVISION?

Restructuring Your Prime Time

Dale and Karen Mason

ACCENT BOOKS
Denver, Colorado

All Scripture references taken from the *Holy Bible, New International Version*. Copyright © 1973, 1978, 1984 International Bible Society. Used by permission of Zondervan Bible Publishers.

ACCENT BOOKS

A division of Accent Publications, Inc.
12100 West Sixth Avenue
P.O. Box 15337
Denver, Colorado 80215

Library of Congress Catalog Card Number 89-82450

ISBN 0-89636-265-5

DEDICATION

We lovingly dedicate this book to our four little blessings, our children: April, Kristin, Analisa, and Taylor, who have taught us more about applying God's Word than we ever thought possible from ones so young.

Acknowledgments

We owe a great debt of gratitude to our family and friends who advised and encouraged us, and to those who selflessly provided babysitting for our children when it was needed most to enable us to continue our work without interruptions. We thank our prayer warriors for their faithful intercession for us on a regular basis. Most of all, we praise our great and mighty God who meets our needs each step of the way—to Him be the glory!

CONTENTS

Introduction 7

PART I - OUR STORY
1 - Our Story 9

PART II - MISUSING THE MEDIA
2 - Tne Non-"Communicable" Disease 14
3 - The Family Altar - Redefined 22
4 - Hollywood's Hidden Agenda 30
5 - Seven Good Reasons to Watch What
They Watch 44

PART III - GAINING CONTROL - A SYSTEM THAT WORKS!
6 - The Program Evaluation Form 57
7 - How Much Do You Really Watch? 70
8 - Withdrawal Symptoms Survival Guide 76
9 - VCR: Friend or Foe? 90

PART IV - TV ALTERNATIVES FOR CHRISTIAN FAMILIES
10 - Changing Old Habits 97
11 - TV Alternatives 102

PART V - THE VIDEOGUIDE FOR CHRISTIAN FAMILIES
12 - The Videoguide 189

Video Index 265

INTRODUCTION

We wrote *Married to Television?* as a husband and wife team. Each of us relied heavily on the other for advice and assistance in writing our respective parts of the book. However, for clarity, we wish to explain that Parts I, II, III, and V are Dale's portion. Part IV, regarding alternatives to TV, was written by Karen.

We have thought of you, our readers, many times as we wrote *Married to Television?*. So often we have prayed that God would enable us to write and then use our words to help you and/or your family. We encourage you, also, to bow in prayer before you begin reading. May God use something in this book to better equip you to live and serve as He desires.

God Bless You!

Dale and Karen Mason

Notice

as possible:

Due Date: 2/23/2006

Stanley J. Grenz & Roger E. Olson.

01 G866w, 1996

reases the longer you keep the item.
is not returned.

CHAPTER ONE

OUR STORY

". . .Don't worry if you've suffered a bankruptcy. And don't be discouraged if all the other car dealers tell you that you have no credit. At Sonny J's Auto Mart poor credit, even bankruptcy, is no problem. . . ." A used car commercial flashed across the screen, interrupting the detective show that I had already seen once or twice before.

I picked up the remote control and hit the mute button. Lowering my size twelve feet from their perch on the coffee table, I rose from my well-warmed dent in the couch, walked the few steps across our small apartment, and quietly opened the door to our bedroom.

It was after midnight, and I wanted to be careful not to wake my wife. But to my surprise, in the dim, dancing television light that slipped through the half-open door, I could see Karen kneeling beside our bed, hear her pouring out her heart to God. She was so deeply involved in her prayer that she was not aware that I had begun to enter. She did not know that I could hear her heartbreaking petition.

My wife was praying for me. She wasn't praying for my health. Nor was she praying for the requests that I had shared throughout the day. No, with quiet sobs she was pleading, "Dear Lord, please help our family. We're hurting. We don't have our husband and father like we need him. Help him to *want* to give up this bad habit. There's nothing more I can say to make him change. It only makes things worse. You have to do it, Lord. God, I'm trusting you to help us. We need Dale. Please don't let him waste our lives by the way he watches television!"

I quietly backed out of the room and gently closed the bedroom door. Bewildered, I stood pondering what I had just heard. After a moment, I walked the few steps back to the couch and sat staring at the silent, moving pictures. "Is she really serious?" I asked myself. "I enjoy watching TV, yes, but I'm not that much different than any average guy. I watch the same basic programs as everyone else." But the prayer echoed in my mind and slowly began to touch my heart.

ADVICE UNHEEDED

I thought back to a day in a Marriage and Family college course when an amiable professor had shared his own bits of married wisdom. The experiences that he communicated so easily held the attention of an idealistic group of soon-to-be-unleashed newlyweds. With an air that suggested the wisdom of experience, he voiced a deeply held personal conviction that now filled my mind again. He entreated, "Don't have a television in your home during your first year of marriage. Establish communication and closeness without the strains that a poorly managed TV can bring."

I had fully intended to follow that advice. In fact, my fiance and I had even discussed the idea after class that day and determined to take his advice. We would be TV-free for one full year. For Karen, this decision was a very welcome, and I believe, a very easy one. She was used to life without TV. But for me, it was a cautious, somewhat timid decision. My habitual viewing was more than relaxation. It was a way of life whose roots extended back to my early childhood.

Like most youths, I longed to hear the loud ringing of the final school bell of the day. I would run home, raid the kitchen pantry, and then flop down on the family room floor in front of the television for an afternoon of kids' programs. Both my first and last thoughts of the day were molded by TV. And as I "matured," the hours I spent in front of the television shifted to a late night concentration. Over the years I graduated from *Leave It To Beaver* to *Happy Days*. My infatuation for *Captain Kangaroo* switched to *Star Trek's* Captain Kirk. My early college days even found me addicted to a daily intake of one of the many sleazy afternoon soap operas.

For me, a decision not to purchase a television for at least one full year was a very big one. But the idea seemed good, and I agreed. Besides, we didn't have enough cash to purchase a set anyway.

So, Karen and I entered wedded bliss without the most standard of entertainment appliances. We returned from our honeymoon to a cozy little apartment free of the sounds of sirens, gunshots, canned laughter, and referees' whistles. We enjoyed evening walks and bicycle rides. We invited friends over to play board games at our dining room table, and we spent time reading God's Word and praying. I was enjoying activities that I had never given serious thought to before. But I was aching, also. I remember looking at my watch at around 10:30 each night and craving to tune into *M*A*S*H* or a late-night talk show. I missed my old "friends," but I was genuinely enjoying a newfound sense of freedom.

A WELCOME VISIT

Enter one very well-meaning widow. About six weeks after Karen and I had set up home on our own, we received a visit from Karen's mother. She had driven 200 miles to our small, college-town apartment. She carried with her some belated wedding gifts, miscellaneous household items, and a strong desire to please her new son-in-law.

She didn't know of our decision to start our marriage without TV. Therefore, she couldn't appreciate the mental gymnastics taking place in my mind when she enthusiastically offered to buy us a television set as a wedding gift. I reasoned to myself, *Well, I've got a pretty good handle on my TV habit now, I'll be okay. Karen won't want it! She's not much of a TV viewer. But this is her mom. I know, I'll tell Karen that I feel obligated to accept the offer, so as to avoid any hard feelings. Besides, I'll only watch the news.*

"Wow! That'd be great," I responded, trying to subdue my enthusiasm enough to be able to build a credible story to Karen once her mom had gone. But that decision—to add a TV to our home without ever having been trained to use it properly—now haunted me. Now the woman that I had married only two years before was at that very moment requesting God's intervention to help realign my priorities. But even as she knelt on the other side of the apartment wall less than ten feet from me, God was answering her prayer!

HOW GOD WORKS

It was because God had allowed me to see the woman that I dearly loved, distressed and at wits' end, that I seriously began to re-evaluate the use of my time. I didn't tell Karen that her words had been heard by someone in addition to God. My pride wouldn't allow me to do that. But with her prayer echoing in my mind, I began to seek God's help and to try earnestly to find alternative activities in which to involve myself. I sorely wanted to impress Karen with actions that would prove that my free time could be spent on something other than watching television.

As I began to pursue non-TV activities, I became much more aware of articles that reported or explored the subject of modern society's use of television. As I read, and as Karen and I observed the habits and attitudes of friends and relatives, we became increasingly alarmed and deeply concerned for the millions of other Christian families who—knowingly or not—were in a situation very similar to ours.

My desire to delve further into this subject intensified. I read everything that I could get my hands on that discussed television and related topics: newspaper articles, magazines, books, research papers, newsletters—all these were fair game to my searching eyes. Now, rather than staring blankly at a glass screen all evening, I was researching the very subject that I was also struggling with. God was using the materials I was reading to cultivate within me the desire to change. He was showing me, indisputably, that uncontrolled and untrained use of TV can and often does have distressing consequences. And the more I discovered, the more I yearned to share the information with others.

One fact that served both to comfort and concern me was the realization of exactly how similar my habits were compared to all the other millions of TV-viewing Americans. I watched basically the same amount, and basically the same types, of shows as the rest of North America. But I also discovered that most viewers, including the overwhelming majority of Christian homemakers, factory workers, professionals, ministers, and business executives, are willing failures when it comes to the issue of TV management.

CHAPTER TWO

THE NON-"COMMUNICABLE" DISEASE—TV ADDICTION

Karen was opening another small packet of soup crackers and handing the two square saltines to one of our anxious toddlers as I charted a course back to our table. Balancing my second helping of lettuce and fixings from the salad bar, I dodged a hurrying young waitress and paused momentarily as a busboy's load of dirty dishes went rolling by.

As I laid the chilled salad plate on the table and pulled out my chair to sit down, I noticed a cozy table for two to my right. Through the branches of an artificial tree, I could see a husband and wife who were out together for Sunday dinner. However, it was obvious that the husband enjoyed Sunday afternoon football more than the company of his spouse. Laying between them on the table was his miniature, two-inch TV, a self-imposed barrier to any conversation. Oblivious to his surroundings, he sat with an earphone crammed in one ear and a finger in the other. His sullen wife stared out a window waiting for the next "time-

14

out" when she could again have 60 seconds to try to share a feeling and hope for a spark of concern. It was clear that watching TV had taken priority over communication in their marriage.

For many viewers, male and female, TV has a damaging hold on their lives. We tend to sit so close to our little screens that we aren't able to see the big picture. Few viewers admit—or recognize—that they have lost control. Only a *very* small minority seem to be struggling with the issue at all.

This chapter presents the statistics, examines the price we pay for our TV habits, and begs us to shoulder the responsibility for our own misuse of television. It encourages us to take a few steps back and look at our lives. Sadly, most Christians have seemingly failed the entrance exam to TV Management 101.

THE TV HABIT AND THE CHRISTIAN

I wonder if we realize how much television has changed our culture? We have an entire generation of young people who choose TV (or video games) almost unanimously over other after-school and evening activities; young people who have never been taught how to find satisfying alternatives and who seem to have all but forgotten how to interact with families and friends.

The daily routine of passive TV viewing monopolizes the free time of both children and adults. At the very least, indiscriminate television viewing is a bad habit. For all too many Christians, it is a very real addiction. Let's face it. How often do we settle down in the evening and breeze through the channels until we find what appears to be the most entertaining program—regardless of the content or morals displayed?

Unfortunately, the typical Christian attends church

faithfully, singing with commitment the desire to have Christ take his life and let it be committed to the Lord Jesus, then returns home to turn on the TV and plop down to share the same vicarious sex-filled, violent, and often profane experiences as his non-Christian neighbors.

As Christians are naively sucked into responding on cue to pre-recorded laugh tracks and crude talk-show hosts, conversation and commitment seem to be forgotten. Or maybe, they were never really learned by the generation labeled "baby boomers" and the children of baby boomers.

Although startling evidence continues to be released regarding the negative effects of indiscriminate and/or large doses of TV, few of us have made personal application of the abundant advice offered by godly experts. There is an alarming degree of spiritual indifference in regard to television. Seldom does a family discuss how the on-screen action contradicts the biblical norms by which we are instructed to live, nor do we realize the number of hours TV is viewed.

Unfortunately, neither has there been any significant improvement in quality control efforts to determine what shows are allowed into our homes. Rather, control of TV content has been entrusted to a small, elite group of network executives whose power to program the thought patterns of an entire nation from lavishly furnished suites in New York and Los Angeles is fueled by enormous sums of money, godless desires, and the love of power.

In most households, the TV schedule dictates the evening agenda, rather than a careful scheduling of preselected programs or non-TV activities. Often, without even questioning the content of the shows that are about to enter our living rooms, we zap a bag of microwave popcorn, pull the tab on a cold diet soda, and place both body and mind in front of the nearest television set.

WHAT THE STATISTICS REVEAL

At first glance, Americans appear to be busy—so busy that hardly another activity could be jammed into our schedules. Some of the many things that fill our week include: full and part-time employment, church services and activities, sporting events, bill paying, schoolwork, housework, yardwork, grocery shopping, telephone conversations, auto maintenance, food preparation, eating, sleeping, personal hygiene, reading newspapers, magazines, *et cetera*. However, recent surveys reveal that the average American household still finds time to tune in the TV for "50 hours each week—more than a normal work-week"[1] or an average of more than three hours per person per day.[2] While most of us complain about the increasing amount of violence and immorality portrayed on television, we "stay tuned" nonetheless.

The immediate gratification box has become an indispensable appliance in American households. We strategically place TVs throughout our homes for ease of use and undisturbed viewing and arrange our living room furniture, our mealtimes, bedtimes, even our bathroom breaks around television. In fact, it appears that we find television more essential than indoor toilets! Televisions are found in 98 percent of American households, while indoor toilets have been installed in only 97 percent.[3]

Recently, while stopped at a busy intersection, I noticed a young couple in a small car pulled up next to mine. They weren't paying any attention to each other; neither did they seem to be aware of the beautiful sunset blazing across the western horizon. Then I noticed that both occupants had their eyes fixed on a small object resting on the console between them. Plugged into the cigarette lighter and glowing in the dim light of evening was a portable black and white TV. Not only a bad idea, but watching TV while

17

driving is illegal as well.

It seems that the more programming that becomes available, the more we feel obligated to view as much of it as possible. The easier it is to transport a TV with us during our daily activities, the harder it becomes to say "no" to its availability. . .or its insidious impact on our lives and attitudes.

Sadly, many parents choose to buy a second (third or fourth!) set, hoping to buy some peace by avoiding some of the inevitable irritations that arise from disagreements about what will be watched. Rather than exerting control and responsibility, they take the path of least resistance.

An evening stroll through almost any neighborhood reveals that, in many homes, parents watch TV in the living room while the kids watch, unsupervised, in their own bedrooms, the kitchen, or the family room. And, as a natural result, the more popular "personal" television sets become, the less we engage in meaningful, much needed conversation. Instead, we march to our designated section of the house and allow the writers and producers of network and cable programs to dictate our thought patterns and shape our world view.

While well-meaning parents may be "keeping the peace," they are also forfeiting precious, never-to-be-repeated opportunities to nurture lifelong family relationships. What we must ask ourselves is: "What price are we paying for our TV-infatuation?"

THE COST

For more than four decades television has been *mis*-represented as one of the least costly forms of personal and family entertainment. Its proponents argue that, once a set is purchased and planted in its own little corner, it costs only about 8 cents[4] per day for the electricity required to power it. While TV may be one of the least costly forms of entertainment in monetary terms, it is one of *the most* costly entertainment choices in terms of opportunities lost.

Within a generation of the creation of commercial television, the medium has deeply entrenched itself into our lives. All-too-often we forget that when we choose to watch TV, we are also choosing *not* to do something else. We have become calloused to the increasingly amoral, anti-God programming. Think about all the centuries before 1950. People did not sit idly in their cottages or castles waiting for television images to appear. No. They filled their lives with activities and people. As the hours of viewing have multiplied, though, the time and attention left for people have vastly diminished. Too often we become guilty of child neglect, spouse neglect, friend neglect—God neglect.

Consider your supper/dinner hour. It used to be a time for the family to visit. But how often is this important portion of the day accompanied by watching TV? Joan Anderson Wilkins in "Breaking the TV Habit" (*Readers Digest*, October 1987) said, "Once upon a time, television was a dessert, something couples shared after dinner, after daily chores, after talk time. But today...it has become the whole meal, and benumbed silence has replaced loving conversation."

The cost of TV? High indeed! We are paying for an electronic sedative that is voluntarily and habitually injected into both adults and children. Its side effects: The erosion of personal and spiritual relationships and the thoughtless waste of a priceless, non-renewable, supernatural resource—time. Where something else once was, television now is. Even if everything funnelled into our homes via the television was "good," we would still need to weigh carefully the merits of sitting idly in front of the screen against the other activities that TV replaces.

19

THE BUCK STOPS HERE

It is important to realize that the problems associated with the use of TV are not directly caused by the electronic invention itself. A TV set does not require our full allegiance from the day that we bring it home. It doesn't yell across the room and sternly demand that we give it the majority of our free time. Its presence only *suggests* that we begin watching. And we do! Periodically, we begin to stay up too late watching TV on Saturday nights. We begin to choose network programming over evening church programs. When unsupervised, children begin to pay just a little more attention to afternoon reruns than their schoolwork, a good book, household chores, or even active outdoor play. And, of course, adults happily accept the never-ceasing assertions that we deserve a break, looking to TV and its mindless game shows, soap operas, sporting events, and situation comedies to relax us from the tensions of the day.

However, we cannot cast all the blame for the problems that arise from immoral television programs upon unregenerate media executives. After all, no one forces us to watch the programs they present. While secular writers and producers must certainly accept responsibility for their large contribution to the moral decline of America, Christians have been guilty as well. Ted Baehr, a member of the board of directors of the National Religious Broadcasters, contends in his book, *The Movie and Video Guide for Christian Families*, that the "anti-church, anti-American, anti-everything attitude prevails in Hollywood today because the churches have retreated." In many ways our passivity and lack of interest has been as much of a contributing factor as that of Hollywood's continually degenerating moral code. We still sit down and view the questionable programs. We still purchase the products that make the shows profitable. Responsibility for the problems that arise from violent, immoral television programs rests squarely upon the shoulders of those who habitually flip the "on" button as much as it does upon the program providers themselves.

The vast majority of us no longer benefit from the positive aspects that carefully selected TV programs and video cassettes can provide. We have slowly fallen into the trap of indiscriminate, overuse of TV. Whether a bad habit or an actual addiction, we need help.

There is no better time than right now to reconsider our relationship to TV and commit to a fresh, new stance based on consistent Christian principles. The best first step that I can think of is to discover what God's Word says in relation to our TV viewing habits.

FOOTNOTES:
[1] Joshua Meyrowitz, "The 1-Inch Neighborhood." *Newsweek* (July 22, 1985): p. 8.
[2] Kevin Perrotta, *Taming The TV Habit*, (Ann Arbor, Servant Publications, 1982), p. 15.
[3] 107th Annual Statistical Abstract, U.S. Bureau of Vital Statistics.
[4] Conversation with SALT RIVER PROJECT Customer Service Department, September 1987, Mesa, Arizona. Cost listed is based upon a modern color television using 80 watts of electricity per hour, seven hours of use per day.

CHAPTER THREE

THE FAMILY ALTAR— REDEFINED

The TV is my shepherd. My spiritual growth shall want. It maketh me to sit down and do nothing for His name's sake, because it requireth all my spare time. It keepeth me from doing my duty as a Christian because it presenteth so many good shows that I must see.

It restoreth my knowledge of the things of the world and keepeth me from the study of God's Word. It leadeth me in the paths of failing to attend the evening church services and doing nothing for the kingdom of God. Yet, though I shall live to be a hundred, I shall keep viewing my TV as long as it will work, for it is my closest friend. Its sounds and its pictures they comfort me.

It presenteth entertainment before me and keepeth me from doing important things with my family. It fills my head with ideas which differ from those in the Word of God.

Surely no good thing will come of my life because of so many wasted hours, and I shall dwell in my remorse and regrets forever.[1]

- Author Unknown -

American Christians are, for the most part, no longer people of the Book. Surveys indicate a disappointing lack of knowledge about our faith and recent news reports reveal an even more disappointing lack of adherence to that faith. Many of those who were once thrilled by a vibrant relationship with Jesus Christ have allowed a hunger for the Bread of Life to be replaced by a hunger for the visual smorgasbord of television. Somewhere there lies a balance that avoids "television gluttony," but chances are, it won't be found while sitting there with the TV control unit in your palm.

Let's take a few minutes to discover some of what the Bible has to say with regard to television.

MY SPIRITUAL GROWTH SHALL WANT

Ever since that life-changing era in the late 1940s when TV sets first began to appear in large numbers on retail store shelves, Christians have had to decide how to deal with this marvelous yet mesmerizing communication tool. The simple solution, "Don't purchase a set," proved to be no solution at all. Media-abstinence was a short-lived option for most Christians. Surveys in 1953 showed that TV ownership was approximately the same percentage between evangelicals and the general public.

By the mid-1960s, TV ownership had become so prevalent that MOODY MONTHLY included an article which stated:

The local church is undoubtedly affected in a definite way by TV competition. Many believe that excessive Saturday night viewing makes an appreciable difference in attendance at Sunday school and church on Sunday morning. Certainly the Sunday evening service finds itself competing with one of the most glittering entertainment arrays of the week. Midweek

23

activities likewise suffer from the pull of the easy chair and the television set.[2]

And in the 1990s, TV gets most of our prime time. It's just so easy.

One of the clearest of Christ's instructions is found in Mark 12:30. "You shall love the Lord your God with all your heart, and with all your soul, and with all your mind, and with all your strength" (NASB). In other words, we are to cling to and give ourselves *wholly* to God. We are to place Him above all else, submitting our thoughts, desires, and goals to His loving care and the principles and commands of His Word. If our lives are truly to be lived according to His plan, we must learn to see the world as God sees it. The only way that we can gain God's perspective is to filter our television choices through His guidebook, the Bible.

SCRIPTURAL DIRECTIVES

In spite of the fact that television wasn't developed until the 1900s, God's Word is far from silent with regard to the use of television. The Bible's clear guidelines relate either directly or indirectly to all areas of everyday life. Our use of television is a small barometer reflecting our personal application or disobedience of these guidelines.

For those who doubt that Scripture has anything to say about mastering this modern assemblage of wires and circuitboards, consider the following passages with your favorite programs in mind.

- *...I will walk in my house with blameless heart. I will set before my eyes no vile thing.*

  (Psalm 101:2b-3)

- *Be imitators of God, . . .among you there must not be even a hint of sexual immorality, or of any kind of impurity, or of greed, because these are improper for God's holy people. Nor should there be obscenity, foolish talk or coarse joking, which are out of place, but rather thanksgiving. . . . Have nothing to do with the fruitless deeds of darkness, but rather expose them. . . . Be very careful, then, how you live—not as unwise but as wise.*

(Ephesians 5:1,3-4,11,15)

- *For the sinful nature desires what is contrary to the Spirit, and the Spirit what is contrary to the sinful nature.* (Galatians 5:17)

- *The acts of the sinful nature are obvious: sexual immorality, impurity and debauchery; idolatry and witchcraft; hatred, discord, jealousy, fits of rage, selfish ambition, dissensions, factions and envy; drunkenness, orgies, and the like. I warn you, as I did before, that those who live like this will not inherit the kingdom of God. Those who belong to Christ Jesus have crucified the sinful nature with its passions and desires. Since we live by the Spirit, let us keep in step with the Spirit.*

(Galatians 5:19-21,24-25)

- *Therefore do not be foolish, but understand what the Lord's will is.* (Ephesians 5:17)

- *Finally, brothers, whatever is true, whatever is noble, whatever is right, whatever is pure, whatever is lovely, whatever is admirable—if anything is excellent or praiseworthy—think about such things.*

(Philippians 4:8)

The hours spent with television are not simply hours of easy and well-earned relaxation. They are not even just a relative waste of time. They are the terrain on which a spiritual battle is fought. Hour by hour, sitcom by soap opera, we are choosing whom we will serve. Will we demonstrate our love and obedience to the God who created us, or will we submit ourselves to the seductive glow of the lesser god?

MODERATION vs. OVERLOAD

None of this means that television, the medium, is inherently bad. *When used in moderation*, it can be an educational and cultural benefit. One study indicated that children who watch *some* TV (about one supervised hour per day) seem to do slightly better in school than children who never watch it. "Actual medical studies, done separately at Stanford University and the University of Southern California, show that programs like television situation comedies can have amazing health benefits. . . .The research indicated that watching comedy shows. . .can reduce harmful stress, relax muscles, stimulate the heart and respiratory system and improve circulation."[3]

Also, certain groups within our society have a greater "need" for the benefits that TV can bring. Elderly and homebound men and women look forward with great anticipation to family-oriented programs, since they often provide a feeling of companionship. (Of course that may mean some of us should just turn off *our* TVs and visit these folks occasionally.) Single parents left alone to fill the roles of mother, father, chef, chauffeur, spiritual leader, playmate, breadwinner, homemaker, disciplinarian, nurse, and auto mechanic find the ability of TV to occupy the kids temporarily to be a much-appreciated asset. Used in moderation and with keen discernment, television can provide the overex-

26

tended single parent with some free time during which she (or he) can retreat and attempt to accomplish some of her many tasks.

Unfortunately, Americans are not exactly famous for moderation or self-discipline.

"When all the hours we've spent watching TV are totaled up against those we have spent in God's Word, I'm sure most of us will stand guilty of allowing this electronic distraction to rob us of one of the most important priorities in our lives."[4]

It has been estimated that the average reader can easily read through the Bible in only 70 hours. What this means is that if the average TV viewer, who watches 3 hours per day, would substitute just 24 days of his TV diet with Scripture reading, he could read the entire Bible. Even slower or more studious readers would have no trouble getting from Genesis to Revelation in double that amount of time. But there is one catch. Discipline. The viewer must resolve to temporarily replace some or all of his TV viewing with a different kind of input.

An even easier challenge would be to cut your average daily intake of TV by just one-sixth, to no more than two-and-a-half hours a day. At the same time, schedule no less than thirty minutes per day for Bible reading. By doing so, an entire Bible read-through can be completed in less than 20 weeks. Better use of future free time is an almsot certain by-product.

The problem is that Bible reading, prayer, and the application of God's Word require effort. Why choose an activity that requires effort when we can grab a soda and the remote control unit, settle back into an overstuffed couch, and have an evening in front of the TV? God's "still small voice" can be conveniently drowned out by the chase scenes, rude remarks, shoot-outs, and sexual liaisons of an average TV evening.

A PERSONAL APPRAISAL

Look at the world around you. Do doctors develop and maintain proficiency at practicing medicine simply by attending an occasional seminar? Or by graduating from years of medical school? No. On the contrary, they practice what they have been taught, and lifelong study to remain current is required. In the same way, to develop a personal relationship with Jesus Christ, we must dedicate significant periods of time to serious study and application (practice) of the Scriptures.

What about you? Is there time for serious, uninterrupted prayer and praise in your life? Or are you among the millions of Christians who invest one or two hours a week in worship and personal Bible study and 20 or more in front of the TV? Does the spiritual climate in your home more closely resemble the warm, lushly growing tropics of South America or the barren lifelessness of Antarctica? Has infatuation with the TV set replaced more important priorities in your everyday life? Has your television been elevated to the position of a new type of "Family Altar"?

We can't be naive. Our adversary is extremely cunning. Satan has been very successful in his unceasing efforts to titillate us with the things of this world, and TV, in its very proliferation, is perhaps one of his best misused snares in modern society.

While we cannot become recluses, simply sitting in our bedrooms reading the Bible all day, neither can we ignore the need for godly, responsible television management.

Our bodies are the temple of the Holy Spirit (I Corinthians 6:19-20). As such, we have a responsibility not to conform to the world (Romans 12:1-2), but to be different, to use different standards for that which we allow into our minds, our hearts, our lives.

Whether you give too much time and attention to the television, a set of well-used golf clubs, or even a pair of lightning-fast knitting needles, remember: too much of even a good activity can be detrimental in the long run—if that activity comes between us and our spiritual health.

FOOTNOTES:
[1]"The 23rd Channel," *Kentucky United Methodist Church* newsletter (March 18, 1977); as quoted by Gregg Lewis in his book *Telegarbage* (Nashville: Thomas Nelson Publishers, 1977), p. 149.
[2]M. Larson, "The One-Eyed Giant." *Moody Monthly* (October 1966); p. 26.
[3]John W. Bachman, *Media - Wasteland Or Wonderland*, (Minneapolis: Augsburg Publishing House, 1984), p. 33.
[4]Bob Maddux, *Fantasy Explosion*, (Ventura, California: Regal Books, 1986), p. 92.

CHAPTER FOUR

HOLLYWOOD'S HIDDEN AGENDA

THE BLURRING OF TRUTH

We were angry, disappointed, and frustrated. Although we had only seen a few episodes of a new "family" TV show, our guard had been let down quickly. The prime-time production was superbly written, a stirring combination of out-of-the-ordinary characters and intellectually arousing situations. In this episode, the plainly dressed Amish family was having some difficulty in adjusting to the modern technologies and social differences of "civilized" America, but their strong reliance upon and devotion to God was often very evident. As a general rule, the writers of this show had chosen to treat God fairly. He was not mocked. In fact, He was held in very high regard.

Recently, though, things had been changing. The previous week we had watched as religious concepts of the humanistic New Age movement were carefully interwoven into the plot. Still, we didn't think that, overall, it was bad enough to cause us to alert the kids. So we just shrugged it off and kept our mouths shut.

This episode, however, this blatant attack on biblical morality, could not go unheeded. A premarital act of fornication between one of the female stars (a recently widowed young woman) and a man whom she had known barely more than one week was, at first, met with righteous anger by the young widow's mother-in-law. We watched anxiously to see how this woman of God would help the lonely, beloved daughter-in-law to acknowledge and ask forgiveness of God for her sin.

But it was here that the story took an alarming twist. The widowed girl's non-Amish mother encouraged her immoral choice. In a brief confrontation with her Amish mother-in-law, the young widow defiantly stated that she was not about to apologize to anyone. "I believe in what I did" (sex with a practical stranger). She accuses her Bible-believing mother-in-law (and representatively, all Christians) of being unfeeling, rigid, and judgmental. The Amish mother-in-law then quickly makes an unrealistic 180-degree turnaround by admitting the error of her ways and her need to "adjust" to a changing moral climate. The show ended with loving hugs and warm-hearted acceptance of the girl's sin.

Apparently, if two people are intelligent and caring, they will put aside their outdated ideas—and the Book in which they are founded—and quickly accept this and other similar "social differences." The Hollywood writers of this particular show even went so far as to parallel the acceptance of premarital sex with progressing from the horse and buggy to the automobile.[1]

SEX EDUCATION 101: HOLLYWOOD STYLE

Television is probably the greatest sex mis-educator in America today. Beware! Network TV would have us believe that sex is simply an activity to indulge in whenever, wherever, and with whomever one gets the urge.

But, sex is God's idea! In the intended boundaries of a

31

loving marriage, it is an exhilarating and creative physical expression of human love. Originating in the paradise of the Garden of Eden, sex is a deeply intimate act created by God to physically unite one man with one woman and to propagate humanity. Designed by God to react with amazing ease to the physical attributes of women, men are especially prone to premarital and extramarital sexual fantasy. Discernment and self-control are most critical for males.

Constantly barraged by the use of sex and sexual connotations as a sales technique, we are tempted many times each day to commit physical and mental adultery. Sex sells! Everything from gossip tabloids at grocery store checkstands, to attention-grabbing highway billboards, to suggestive television and magazine advertisements. This God-given wonder has been prostituted to facilitate the sales goals of corporate America.

Television viewers from cradle to rocking chair are fed a lie that says, "fantasize, enjoy, indulge, no one will ever know!" But God's Word says that sexual fantasy is nothing less than *adultery* (Matthew 5:28). Sexual fantasy has been described by many psychologists and behavioral scientists (not to mention most pastors) as an appetite that is self-perpetuated. Feed it tonight and tomorrow you will hunger for more. Each episode provides a temporary feeling of relief, quickly overshadowed by an even more intense desire for an even heavier dose of visual or physical stimulation.

Greg Lewis, in his book *Telegarbage*, stated, "the beauty of sex, the purpose for which God created it, the guidelines He set for its fullest enjoyment and its powerful potential as the most intimately sensitive communication between a husband and wife are never portrayed on TV." In fact, "references to intercourse on television, whether verbally insinuated or contextually implied, occur between unmarried partners five times as often as married couples; references to intercourse with prostitutes comes in second."[2] The sin of homosexuality is glossed over and presented as nothing more than a state of being that should be accepted as naturally as one's skin color or sexual gender.

AN EVER-CHANGING STANDARD

TV has become the ever-changing standard by which we assess and judge the morality of modern society. One step at a time, it has replaced the Word of God, the original foundation and cornerstone of America, as the absolute guideline for moral choices in our lives. By touching the thought processes of millions of people every evening, "prime time" works to make the values of the coastline liberal elite the values of the nation. Habits, attitudes, and values that traditionally have been instilled by parents, or by the church, are now being taught by television. "It tells us...what is right and wrong, what is acceptable and unacceptable, whom to believe, whom to trust and not trust, and whom we should desire to emulate."[3] Secular television producers are gradually acclimating us to a make-believe world where there are no real consequences for sin. Even the most difficult personal problems can be successfully resolved in 30-60 minutes. Satan, the father of lies, is alive and well in prime time.

Television programming is ample evidence of the truth of Romans 8:5-8, *"Those who live according to the sinful nature have their minds set on what that nature desires; but those who live in accordance with the Spirit have their minds set on what the Spirit desires. The mind of sinful man is death, but the mind controlled by the Spirit is life and peace; the sinful mind is hostile to God. It does not submit to God's law, nor can it do so. Those controlled by the sinful nature cannot please God."*

In an interview with *Contemporary Christian Magazine* several years ago, internationally famous "Star Wars" filmmaker George Lucas candidly confided his opinion that, "the influence of the church, which used to be all powerful, has been usurped by film. Films and television tell us. . .what is right and wrong."[4]

Television has an awesome potential to shape and change almost every aspect of life. We can literally *watch* history change anywhere in the world via satellite transmissions. And, the more a viewer watches, the more pervasive and believable the messages of the media power brokers become. Biblical truths begin to blur. Discernment begins to wane. Actions that were once out-of-bounds begin to be accepted, both on the screen and in society at large. An emotionally stirring scene portraying the victimization of one "nice" gay on television produces 10,000 more in our minds. No blatant speech by a pro-homosexual actor is necessary. These attitudes need only be suggested. Before long, our reactions are deadened and behaviors which we once found wrong are no longer perceived as such. We no longer remember that television programs are not a true-to-life, balanced view of how the world really is.

A HIDDEN AGENDA?

In his revealing book, *The Home Invaders*, Donald Wildmon documents the findings of the research team of Lichter and Rothman. They conducted in-depth survey/interviews with 104 of the most influential professionals in the television industry. Among those interviewed were 15 presidents of independent production companies, 61 producers (26 of whom are also writers), and 10 network vice-presidents responsible for program development and selection. These individuals were identified by Lichter-Roth-

34

man as "the cream of the television creative community. . . some of the most experienced and respected members of the craft." However, many alarming statistics about the prejudices of the people who decide what we are offered were revealed. Among them:

* 93% seldom or never attend worship services.
* 97% feel that the pregnant woman carrying an unborn child has the right to decide on abortion.
* A mere 5% strongly agree that homosexuality is wrong.
* Only 16% strongly feel that adultery is wrong.
* An astounding 99% are of the opinion that TV programs should be a little more critical of traditional values.

Secular television's architects, biased against Christianity and its essential moral guidelines, admit that they are not in it just for the money. They are trying to move their viewers toward their own ideal of the good society.[5]

This group of television executives plays a major role in creating and shaping the programs whose themes and stars become idolized in our popular culture. In an article from the *American Family Association Journal*, Judy Price, Vice-President in charge of Children's Programming at CBS, stated, "I think we've broken a lot of ground (in children's programs) where people would not have dared to go in prime-time." When questioned as to her motivation for taking control of the Children's Programming division she responded, "I could get away with more."[6]

Christian adults must ask themselves: "Are these the people I want to shape my thoughts, my child's attitudes?"

ORDINARY LIVING

Have you ever noticed what is left out of the pseudo-world of TV programming? We are entertained by lifestyles in which the ordinaries of living are absent. Gone are the daily routines, the self-sacrificial actions and healthy relationships exemplified in the lives of millions of real people. We are given, instead, a glamorous, illusionary world, which we welcome for its vicarious experiences. But we retire from our daily viewing with a new collection of glittery, unrealistic expectations to which we can never attain.

CREATOR-LESS

What else is missing from the TV screen? What topics and stories should we encourage the network and local TV stations to dedicate air-time to? How about a decent individual with a sincere, biblically-based belief in God? For the most part, the only people on secular television that believe in God are kooks and fanatics. True Christians find themselves portrayed as oddities or relics from an era long past. It's no wonder our neighbors have such a distorted view of Christianity!

Belief in God has been edited, redefined, and replaced. By emphasizing the supposed self-sufficiency of man, TV writers and producers train us to ignore the hand of God, to deny His rightful authority over our lives. To put it simply, the television world view ignores God.

Take this logic a step further, and one could safely say that television also teaches us through what it does not tell us. By what is left out of the programming, it teaches that the Creator does not exist and therefore His purposes and expectations are not important. In *Taming the TV Habit*, Kevin Perotta states, "No made-for-television movie leaves the viewer marveling at how God worked out everything for the good of the characters who trusted Him. No television hero, in a moment of humility, admits his inability to right an injustice and calls on God to act. No news commentator reflects on the rise and fall of nations in light of the Biblical prophets who spoke about the kinds of behavior which God rewards and punishes."[7] Created man, not man's Creator, fills the screen. Can you imagine your favorite program with a biblical basis, a godly world view?

The television view of the world completely excludes an awareness of God. This ploy of Satan distracts people from thinking about who God really is by filling our minds with man-centered stories and man-centered explanations for God's supernatural acts. Even the Corporation for Public Broadcasting (PBS/NPR)—underwritten in large part by all American taxpayers[8]—unwaveringly promotes the theory of evolution, as if it were proven fact! Biased television producers effectively withhold from both news and drama the wealth of scientific evidences for biblical creation.

It is of practical importance to every Christian to realize that as our Creator, God owns us—and should control what we allow between our ears! We have the mind of Christ (I Corinthians 2:16), and, I'm afraid, that too often what we put inside our minds dishonors Him greatly!

THE INGREDIENTS

Are the ingredients of the normal "TV dinner" fit for human consumption?

Let's start by discussing the appetizer—those all-too-frequent 30- and 60-second commercials that bombard us with images designed to whet our material appetites for every conceivable invention of man. They attempt to seduce us into buying everything from luxury cruises to gourmet dog food by telling us what we are supposed to lust after. If your car is not the latest model, with the smoothest lines, you're "out of it." Even more subtly, we are told that if our spouse is not the most beautiful, with the best body, he or she may not necessarily be worthy of our continued devotion.

In case you are sitting there saying, "Yep. I agree. That's

what commercials do, but I sure don't let it affect me," think again! Why else do highly profitable advertising agencies continue to thrust these types of sales campaigns at us? And why would manufacturers continue to pay hundreds of millions of marketing dollars for these kinds of ads if they weren't accomplishing their intended goal—to entice and arouse us to action? Let's not delude ourselves. We are being trained to covet and to do so unabashedly.

One evening my family and I had opted to watch the TV for a little entertainment. We found a purely wholesome special and were enjoying it immensely. However, a commercial break soon interrupted our relaxation by filling the screen with a provocatively dressed female, slowly caressing and kissing her lover with all the seductive expertise of an uptown prostitute. "Click"—off went the tube until our chosen program returned. This isn't what we're training our pliable little girls, so vulnerable because of their innocence and naivety, to admire and eventually seek to emulate. This isn't what I need to tempt me either. And I knew it certainly was not the kind of appetizer that God expected me to serve to my family as His appointed head of it.

Increasingly, television's main course is sex and violence. Sexual promiscuity, fornication, adultery, and homosexuality are dished up in heaping portions. According to a study released in 1988 by Lou Harris and Associates, Inc., "The three networks broadcast 65,000 references to sexual behavior in the 1987-88 television season, an average of 27 references per hour."[9] And that number continues to rise. Honest now, how often have we caught ourselves heartily joining the canned laughter at humorously portrayed immoralities?

Murders, car crashes, explosions, and shoot-outs are the meat and potatoes of TV's police and detective shows (not to mention the after-school and Saturday morning cartoons).

Violence hardly causes us to raise an eyebrow these days. The more calloused we become through repeated exposure on TV, the more it takes to shock us when the violence is part of the nightly news. We're rapidly becoming desensitized to the point of apathy toward genuine misery and pain.

Then, consider the ever increasing market for horror films and videos. Our neighborhood video shops do a booming business in these shows where Satan and his works are glorified. The customers are not just teenage boys, either. These videos are also rented in large numbers to young girls for overnight slumber parties. To make matters worse, the already lax age/rating guidelines established by the MPAA (Motion Picture Association of America) are not enforced at most video rental counters. Whether the customer is 9 or 29, most store owners seem to operate by the philosophy, "if he's got the dough, he can watch the show!" Today, we have a generation fed by violence which embraces horrors that just a few years ago we could not have imagined being publicly broadcast.

Whether the appetizer, a side order, or the main dish, every item on the television menu is heavily garnished with humanistic selfishness. We are told that we can and should get absolutely everything we want—as soon as we want it. After all, we deserve what we desire. On TV, it is normal to be shamelessly materialistic and self-centered. Scriptwriters have redefined the term sin to the point where an action is only wrong if it obviously hurts someone else, or if the one who commits an illegal act is caught. Guilt, the natural result of wrongdoing, is presented as a mental distress that you can pay a psychiatrist to help you resolve. There is no mention of repentance.

Lust, sexual promiscuity, violence, and selfishness are regular menu items on TV's visual dinner table. The normal network offering is poor nourishment for growing Christians. If anything, it will stunt our growth! No, television is not harmless entertainment that brings unoffending or humorous images into our homes. When we don't carefully select what we watch, it nourishes a non-Christian way of looking at life. From adventures to sports, from soaps to the news, common themes run throughout the content of secular television. It brings, by our own personal invitation, the images of a depraved and wicked world into our minds and hearts.

How many Christians are compromising the guidelines in God's Word for self-gratification and personal convenience? Many of the same parents that demand their children adhere to a rigid nutritional regimen abandon their little (and not-so-little!) ones to consume hour after hour of visual "junk food," oblivious to the harm being done.

WHAT CAN WE DO?

For nearly half a century Christians have urgently needed to monitor what, and how much, is viewed in private on their television sets. However, there are no courses or Sunday School classes on godly TV management. Sadly, the vast majority of us have followed the crowd, putting aside our ability to discriminate, to make conscious, clear-cut decisions on *what* we'll watch, *when*, and for *what* period of time.

Part of the problem may well be that Christians tend to claim some sort of immunity. We falsely assume that because we attend church, television's impurities cannot corrupt us. However, while a few Christians fill a pew every time the churchdoors swing open, most limit their involvement to simply attending Sunday morning services. In fact, the average American Christian spends about fifteen to twenty times as many hours in front of the TV as he does studying his Bible, praying, or listening to his own minister! This imbalance must have an influence on us.

SUNDAY SCHOOL vs. SUPER HEROES

From time to time, my wife serves as a children's Sunday School teacher at our church. I am always amazed as I observe her late nights of preparation. She carefully chooses the lesson concepts. Then she spends hours searching for just the right pictures to illustrate the Bible story. She designs both the craft and song-time to complement the story. All this is done with high hopes that the meager 60 minutes she spends with those impressionable little rascals will leave a lasting mark in the wet concrete of their minds. And she is sure that it does, to varying degrees. However, she winces when considering the many hours of godless images that will capture their attention during the following six days. How much of that gentle footprint of Jesus will still be recognizable after TV's super heroes and super villains have repeatedly stampeded through the wet concrete that she had so little time to shape?

Do we realize that crawling into our Sunday-go-to-meeting duds, packing the kids into the car, and attending Sunday morning services doesn't come anywhere close to guaranteeing a healthy, Christ-centered world-view?

You and I, as God's children, must recognize the awesome power and influence that television has in our lives and in the lives of those around us. TV viewing, which generally seems like an unimportant activity, often used out of boredom, is truly one of the most influential mental battles in which we can choose to participate. Every time we turn to another afternoon or evening soap, a special movie, or even the nightly news, we enter a receptive communion with the images and messages of a blatantly anti-God culture. In the battle of TV viewing, most Christians have chosen to lay down their spiritual armor. We have willingly abandoned our defenses and offered ourselves as unprotected targets for the arrows of the evil one. We have based our viewing decision more on our own personal schedule and habits than on the basis of acceptable program content. And we resist, rationalize, and reject the thought of changing those TV habits.

THE CHALLENGE

The decision to love God with all our heart, mind, soul, and strength has practical implications for our use of television. We must learn to carefully *use* rather than habitually *watch* TV. We must train ourselves to be more discriminating in our program selection. We must be biblically knowledgeable to be able to discern that which honors God from that which shames Him. If there is nothing on that truly merits the investment of our free time, we can ask God to give us the willpower necessary to walk across the room and push the off button. Then we can invest our time in alternative activities that will better serve the Creator's purposes and help us to build a consistent, *Christ*-centered world view. Watching television through boredom or lack of imagination for other things to do denies the creativity God gives each of us as well as the infinite diversity of His gifts to us.

FOOTNOTES:

[1]"Aaron's Way" 1-hour television episode, originally broadcast by NBC (*National Broadcasting Company*, USA) on May 4, 1988.
[2]"Television and Sexual Learning in Childhood" *Television and Behavior* (Volume 2, pp. 209-210, 220-222), quoted in John W. Tranter, Jr., *Images* (Springdale, Pennsylvania, Whitaker House, 1986), p. 69.

[3] Donald Wildmon, *The Home Invaders* (Wheaton, Illinois: Victor Books, 1985), p. 45.

[4] "The Gospel of Lucas," *Contemporary Christian Magazine* (August 1983).

[5] Wildmon, *The Home Invaders*, p. 22.

[6] "CBS Mighty Mouse Sniffing Cocaine?" *American Family Association Journal* (July 1988): p. 4.

[7] Kevin Perrotta, *Taming the TV Habit*, (Ann Arbor, Michigan: Servant Publications, 1982), p. 116.

[8] *A Report to the People*; Annual report (Washington, D.C.: Corporation for Public Broadcasting, 1986), p. f-3 (Statement of Financial Activity).

[9] Larry McShane (Associated Press), "TV Gives Unrealistic View of Sex..." *Mesa Tribune* (January 27, 1988): [This is a summary of a study conducted by Louis Harris and Associates, Inc. and released in January 1988.]

CHAPTER FIVE

Seven Good Reasons To Watch What They Watch: TV's Effects on Kids

This chapter is dedicated to those adults who are not yet convinced of the need to handle television with caution. It is written for those who may argue that, since they watched a lot of television when they were kids and feel that they were not negatively affected, it won't have a significant impact on their children or on them as adults today.

The difference, though, lies in the nature of the programming from "then" to "now." Many of today's baby boomers were brought up on "Leave It to Beaver," "Father Knows Best," "Rin Tin Tin," "Lassie," "Bonanza," "Woody Woodpecker," "The Brady Bunch," and "Underdog." Today we have "Twin Peaks," "L.A. Law," "Captain Power," "Teenage Mutant Ninja Turtles," "Ghostbusters," "The Simpsons," "Cheers," and more. Today's television shows are saturating us and our children with violence, poor morals, materialism, and occultism. Caution is required as well as an unceasing vigilance against our enemy who seeks to devour us, our children, and our witness (I Peter 5:8).

While what you are about to read is a rather long list of the "bad fruits" associated with poor TV management, it is important to remember that *not everything that comes through TV is bad.* Rather, it is overuse and a general, uncaring attitude toward the medium *by adults* that can reap disastrous results.

It is not my intent to alienate you from your TV. In fact, beginning in the next chapter, there are some very refreshing, creative ideas of how you and your family can gain lasting control of and benefit from this valuable communication tool. However, because the average child between 2 and 11 years old watches 27.3 hours of television per week[1]; because "the only thing that kids do more than watch television is sleep"[2], and because I am convinced that most parents are either unaware of, or completely calloused to the indecent liberties that modern media take with our children,*"Your attention please!"*

VIOLENCE

The American Academy of Pediatrics has thoroughly studied the issue of TV violence and its effects on children. Their December 1984 Policy Statement confirmed some long assumed realities, such as: "Repeated exposure to televised violence promotes a proclivity to violence and a passive response to its practice. . . ." Also, the office of the U.S. Surgeon General has investigated the negative effects of television almost as often as it has studied the consequences of cigarette smoking. A *USA Today* magazine article reported, "The government has not insisted yet on labeling television programs with the warning: 'viewing may be dangerous to your mental health,' but that is the

45

inescapable conclusion of massive research."[3]

We have all heard or read a news story about a life-shattering crime committed by some 14-year-old who saw it on TV and thought it would be "fun" to act it out himself. It is probably safe to assume that none of the parents of these young terrorists sat their infants in front of a TV with the intention of training them how to kill and maim, rape and brutalize, but the messages were transmitted, and the children were there to receive them.

Not even a strong Christian home guarantees immunity. Christian parents—who, for the most part, allow the same shows into their homes as do their non-Christian neighbors—need to remember that during a youngster's estimated 22,000 hours in front of television by age 14, he has witnessed the assault on, or destruction of, 18,000 individuals,[4] usually without any negative consequences. Is it any wonder we see children "playing" violently, cursing, or fighting?

The early evening hours when the most family-oriented shows are supposedly on is really the most violent time on weekday network television.[5] According to *USA Today*, violent acts during prime time are surpassed only by the number telecast for the viewing pleasure of impressionable youngsters during Saturday morning cartoons![6]

Psychoanalyst Dr. Jay Martin of the University of Southern California reported that "in a multi-year study of 732 children, conflicts with parents, fighting with peers, and delinquency were correlated with the total number of hours of television viewing." It is troubling to note that the "fundamental correlation is not between aggressive behavior and the viewing of violence on TV. Increases in aggression correlate with *viewing television*, not with viewing violent scenes."[7] The process of viewing—the number of

46

hours actually viewed—is the main factor that causes negative behavior.

It appears that the best way to guard against over-aggressiveness and interpersonal conflicts is a two-pronged approach. First, eliminate all violent programs from your TV-viewing diet. Second, and more importantly, cut down on the total number of hours viewed.

ACADEMICS

Ellen De Franco, in her book *TV On/Off*, said, "Nationwide, reading scores for elementary and secondary children continue to go down: the number of partially illiterate college students is increasing. Classroom teachers are finding it harder each year to establish a healthy learning environment, no matter for what age students they are responsible. It is not only the discipline problem that confronts them, but the imperative to capture and hold children's attention. Teachers complain that children 'turn off' very easily, seem restless and apathetic. Many of the tried and true tricks of motivation no longer are effective. Educators find themselves in competition with television's effects, and they often feel that they are loosing or have already lost the struggle for children's minds."[8]

In a *Parade* magazine article entitled "Why They Excel" (January 21, 1990), author Fox Butterfield discusses the differences in academic achievement between Americans and Asians. Among many other thought provoking statements, she refers to a study prepared for the United States Department of Education that compared the math and science achievements of 24,000 13-year-olds in the USA, Canada, South Korea, Ireland, Great Britain, and Spain. "One of the findings was that the more time students spent watching television, the poorer their performance. The American students watched the most television. They also

47

got the worst scores in math. Only the Irish students and some of the Canadians scored lower in science." An observation shared by a student at Berkeley in this article should cause most parents to seriously reconsider the situation in their own homes. This young Korean-American frankly stated, "I don't think Asians are any smarter. . .there are brilliant Americans in my chemistry class. But the Asian students work harder. I see a lot of wasted potential among Americans."

And recent studies suggest that North America is never going to gain ground in trying to catch up with the academic quality of our overseas neighbors by depending on "educational TV." While most parents assume that "educational" shows are teaching basic skills, and while this is undoubtedly correct to some extent, children who watch these types of shows "tend to solve problems only on the basis of facts or concepts presented. . .whereas children who learn the same materials in a traditional manner solve problems more freely and individually. Decreases in cognition tend to occur whether the program is an adventure show, a comedy, or even an educational program. A broadcast whose subject is how to increase creativity is likely to *decrease* creativity in the viewer."[9]

While headlines in newspapers and presidential studies demand excellence in our classrooms, the evidence that too much television has an adverse effect on scholastic performance is overwhelming. The bottom line is, the more television a child watches, the greater the negative impact on his learning and development. Our educational reforms must be matched by reform in our TV viewing habits. Television impedes children's development of mental habits which they urgently need in order to grow into useful, well-educated citizens, mature human beings, and effective servants of God.

DISTORTED VIEW OF SEX AND SEXUALITY

Adults have watched as the minimum moral standards to which network programs must attain have continually declined. The decline has been great, but slow enough that most Christian adults are no longer shocked or outraged by programs that, only a decade earlier, network programmers would have never dared to transmit. And, as adults become more and more desensitized to casual sex, so are the children that God has entrusted to their care.

After extensive study, the American Academy of Pediatrics, a multi-national organization, found that on television, "sexual relationships develop rapidly; the risk of pregnancy is rarely considered; [and] adolescence is portrayed as a constant state of sexual crisis. . . .Pornography on television is a particularly important concern."[10]

In an effort to broaden and obtain a better selection of programs, a very large percentage of American households now subscribe to cable TV services. But better control is usually the first benefit that cable TV subscribers realize *doesn't* come with their monthly service fee. In fact, in a study of 450 sixth-graders who watch cable, Oklahoma State University professor Godfrey Ellis found that a staggering 66% of the children watched at least one program a month that contained nudity or heavy sexual content.

Where do Christian children develop their weakened moral ideas? A substantial part of the blame can be credited to poorly managed television. A child may attend Sunday School for one hour a week, church for two more hours and never really hear about God's prohibitions regarding premarital sex. But when a child has unlimited access to the world's perspective at the rate of 25 to 30 hours per week, which ideas can we expect to have the most influence?

OBESITY

In a December 1984 policy statement, medical researchers for the American Academy of Pediatrics have found that "television viewing increases consumption of high caloric density snacks and increases the prevalence of obesity."

Results of another study reported in the *Arizona Republic* newspaper stated that, "the proportion of American children who are overweight has increased more than 50 percent over two decades. . . .The implications are that there is going to be a major rise in the prevalence of adult obesity and its consequences."[11] The researchers also noted the lack of physical activity in addition to the increased consumption of high-fat foods as an important culprit.

COMMERCIALS

There are very influential and cunningly deceptive media power brokers out there, hungry in their insatiable desire to garner material wealth. They are obviously willing to sell our kids down a moral and intellectual drain in that process. While television has a lot to answer for, its exploitation of children through commercials may just rank at the top of the list.

It's estimated that the average child sees 20,000 commercials a year. Contrary to adults, who often mute out commercials, or who get up and make a mad dash for the bathroom during the 60 to 180 seconds that they are allotted, children like TV ads. They like to be told what to lobby for. . .and lobby they do. When mom tries to pull a good tasting, healthful box of cereal from the grocery store shelf, her hand is held back by a whining, pleading child who is willing, at that moment, to sell his birthright in exchange for a box of brightly colored, sugar-saturated puffed flakes. The child makes such an embarrassing scene that—although she knows it is the wrong thing to do—the

50

poor mother finally gives in and throws the doubly expensive box and its "free prize" into the cart. It lands on top of a plastic container full of vitamin fortified, fruit-flavored sugar water. Like their parents, our children succumb early in life to the puppet strings of Madison Avenue. They buy into and energetically promote the idea that ability and health are products of material consumption. Commercials are the place where sizzle overwhelms substance and where paid liars can get away with anything as long as they look honest on camera.

For many years, educators have capitalized on the fact that what we see and hear is retained much longer than anything we learn through the use of just one sense. Many of America's best teachers, seminar leaders, and pastors use films and videos, rich with a wide variety of colorful images, background music, and sound effects, to reinforce the material they are covering. Because they do so, the information is absorbed and internalized much better.

Unfortunately, commercials have all the best of advertising minds, multi-media approaches, and frequent repetition going for them. Their lack of intrinsic importance becomes immaterial as these other elements create an overwhelmingly influential message. This is a good case in favor of Christian families building a home library of Christian videos for themselves and their children. At the same time, it is a warning to avoid letting children watch television shows that carry a large number of commercials directed specifically at them, particularly the Saturday morning cartoons, which contain about 25 percent more commercials than other programming.

If you are tired of hearing Junior whine all the way through the grocery store, and then dragging him kicking and screaming from the toy department of the local discount store, turn off Saturday morning network TV. Insert a couple of Christian cartoons, go to the park, or play a game with your children. In the long-run, you have everything to gain and nothing to lose!

IMAGINATIVE PLAY

Few adults are aware of it, but television has completely altered the way that children spend their time. Yesterday's children spent much of their days playing games and exploring the outdoor world around them. But today's children spend their time with their eyes glued to the television screen and their bottoms firmly planted on the living-room rug.

TV has often been identified as a sort of "plug-in drug." This description is really quite accurate. Television gradually narcoticizes viewers into passivity. Youngsters who should be outside getting bruised, dirty, and exhausted, exercise only their blinking eyelids as they sit entranced, hour after hour, in front of the tube. Dr. Paul J. Fink of Thomas Jefferson University in Philadelphia has studied childhood viewing habits and concludes that "those obsessed by TV are less creative and more passive."[12] Evidence also indicates that television interferes with the capacity to entertain oneself and stifles the ability to express ideas logically and sensitively. Television viewing replaces essential play activities with passivity rather than activity. These findings are generally true of adult viewers as well.

A simple observation of children's free play shows that it is often structured closely after TV programs. It seems that every preschooler in the country wants to play with toys that are connected to popular, violent cartoons. Children run around the back yard blasting their friends with imaginary laser-bullets through over-priced plastic weapons that they have learned how to use by watching TV.

Toys linked to television who's program a child to play fashioned after the show—whether violent or benign. Children no longer need to envision situations or new worlds; they simply replay last Saturday's cartoons. Imagination is crippled; inventiveness stunted.

FAMILY TIME

Dad won't be home on time tonight. He's working late at the office. But that's O.K.! Timmy has the TV. The TV is always there to entertain him. Who needs Dad when the TV is in good working condition!?

In his excellent book *Family Issues*, Christian author and radio talk-show host Bob Larson reveals an alarming finding that should make even the most carefree father sit up and think. He says that, "a recent Michigan State University study revealed that when four- and five-year-olds were offered the choice between giving up television or their fathers, a third opted to give up daddy." According to another study, "the average five-year-old spends 25 *minutes* a week in close interaction with his father and 25 *hours* a week in close interaction with the TV set."[13]

Parents often regret their inability to spend more time with their children. However, in a survey conducted for the Massachusetts Mutual Life Insurance Corporation, "two-thirds of those surveyed say they would probably accept a job that required more time away from home if it offered higher income or greater prestige."[14] Caught in time binds that limit the number of hours available for family interaction, equally problematic is the average family's misuse of the TV set.

In a *Saturday Evening Post* article, Marie Winn wrote, "The television set casts its magic spell, freezing speech and action, turning the living into silent statues. . . .Turning on the television set can turn off the process that transforms

53

children into people."[15] Poorly managed television wastes opportunities for kids to learn how to relate to other people—including their parents and siblings. Instead, in the strong words of one author, "Parents have abused their children in order to benefit themselves, turning the TV set into a constant and convenient babysitter."[16]

I'm convinced, however, that the family's loss of control of its time is one of the most volatile problems faced by Christian parents today. Christian parents recognize the fact that values completely contrary to those that they want their children to absorb are being shot rapid fire through the TV set into the living room. They realize that, as the family supper table also succumbs to the chatter of TV noise, hope for a daily period of sharing, caring, and inter-action is almost zero. Yet, they stay "tuned in." When one considers that the average couple now spends almost 50 hours per week with the TV on and only 27-1/2 minutes talking to each other, it is little wonder that relationships between the members of the household suffer. And when families suffer, our entire nation sees and feels the results.

Children of all ages need adult contact. While a teenager's vehement verbal attacks may indicate otherwise, they need adult/child relationships for reassurance that they are loved, and for instruction in the ways of adult society. Author/lecturer Josh McDowell has repeatedly stated that he often has teenagers come to him, convinced that their parents don't love them. When asked why they feel this way, many respond that they just don't feel important. Their parents just don't spend time with them anymore.

Children learn from parental example—whether that example is passivity or loving involvement, harsh words or gentle speech. Teaching children to handle money, to show respect, or to learn responsibility through household chores cannot be sacrificed to the sponge of television programming. Children who do not learn these traits become adults who are selfish, uncooperative, and often lazy.

If someone in your family wants some uninterrupted time to share some problems or feelings, how many times do you respond with a, "Shhh, I'm watching TV"? That phrase is a strong indication that television is the basic presence and all others are considered interruptive. When we put TV ahead of people, it reveals a lot about the value we place on others in our home.

Will you give prime time to your spouse, friends, church? Make a mutual commitment for just one week to change your viewing habits. The remainder of this book is designed to equip you and your family to do just that. It includes very practical suggestions, forms, personal stories, Christian video evaluations, and alternative activities. Are you ready for a more responsible, less habitual use of the "miracle" communications medium, modern television?

FOOTNOTES:

[1] Bob Dart, "Violent Acts Saturate Kids' TV" *Mesa Tribune* (January 26, 1990): p. A6. [This statistic is based on the Neilsen Index.]

[2] Dr. Robert A. Mendelson, M.D., Chairman of the TV Committee of the American Academy of Pediatrics, as quoted in "TV Can Ruin Some Kids." *Journal Star*, Peoria, IL, April 20, 1990: p. A4.

[3] Dr. Jay Martin, Psychoanalyst, "Caught in Fantasyland." *USA TODAY Magazine* (July 1988): p. 93.

[4] J.W. Stein, "Shotgun Wedding: Television and the Schools." *USA TODAY Magazine* (November 1983): p. 431.

[5] Tipper Gore, *Raising PG Kids in an X-Rated Society* Nashville, Abingdon Press, 1987), p. 64.

[6] "Caught in Fantasyland," p. 92.

[7] "Caught in Fantasyland," p. 93.

[8] Ellen De Franco, *TV On / Off* (Santa Monica, Goodyear Publishing, 1980), p. 153.

[9]"Caught in Fantasyland," p. 93.

[10]"Children, Adolescents, and Television." *American Academy of Pediatrics Policy Statement* (December 1984).

[11]"Fatty, Fatty, 2 by 4." *Arizona Republic* (May 3, 1987): p. AA19.

[12]Bob Larson, *Larson's Book of Family Issues* (Wheaton, Tyndale House Publishers, 1986), pp. 254,260.

[13]Robert S. Welch, "Making Your Family #1 in '87." *Focus On The Family* Magazine (January 1987): p. 4.

[14]"Time Bind Called Top Threat to Family." *Moody Monthly* (December 1989): p. 72.

[15]Marie Winn, "The Plug-in Drug," *Saturday Evening Post* (November 1977): pp. 40,41.

[16]De Franco, *TV On/Off*, p. 152.

CHAPTER SIX

THE PROGRAM EVALUATION FORM: *Setting a Standard for Responsible Viewing*

Most parents would probably agree that the innocent questions of a young child can be even more convicting than a face-to-face confrontation with an accusing adult. I have been reminded of this embarrassing fact on several occasions. One particular instance stands out in my mind.

FROM THE MOUTHS OF BABES

It was the weekend. Time to relax. Our family had gone to the home of a relative for an evening visit. After a while, I stole away to the already occupied "TV room." Very soon my two-year-old daughter followed the noise of the television straight to where she knew her daddy could be found. She was wearing one of those "I found you!" kind of expressions. You know, it was the kind of victorious smirk kids exhibit once they have located a cleverly hidden playmate

57

in hide-n-seek. Cuddling up onto my lap and resting her tiny head against my chest, we settled back for a bit of visual entertainment.

Before long we were both fully engrossed in a fast-moving, suspense-filled science fiction movie. About one hour, a dozen curse words and several sexual innuendos into the show, the screen was suddenly filled with a graphic portrayal of torturous murder.

I was caught off guard, but soon remembered the presence of my tender two-year-old. I quickly shielded her eyes to protect her from this gruesome sight. Obviously frustrated, she turned to me and asked, "Why you put your hands over my eyes?!" Determined to maintain my composure I calmly responded—in my authoritative, daddy sort of voice—"Because little girls shouldn't watch bad things like this." But without even a moment's hesitation this little child immediately asked, "Then why *you* watching it?" Convicted by the insightful words and sincere expression of my own young daughter, I responded, rather dumbfounded, "Good question!" Together we rose and left the room.

God used that simple experience to challenge me. My own personal TV viewing selections were often far from wholesome, edifying entertainment. I knew that already. Yet, I also knew that there were positive aspects to some of what I viewed as well. My greatest frustration, though, was with myself. I wanted to be a more consistent example to my wife and children. I earnestly desired to exhibit self-control and to be more discriminating in my program selections. But all the desire in the world did me no good until a simple one-word question could be answered: *"How?"*

SOUND FAMILIAR?

Hopefully, by this point you, too, are pretty well convinced that the TV set inhabiting your home *must* be dealt with. But *how*? Initiating a battle plan requires practical help. Before you load your TV into the family car to hurl it from the nearest cliff, learn to discern helpful entertainment from harmful entrapment. Whether 8, 38, or 68, it's not too late to learn how to master the one-eyed monster.

THE CHALLENGE

In order to change television program content, we must have a more active voice. It is important that viewers occasionally take time to write to program providers. Communicate your appreciation for wholesome shows, express your disgust toward those that mock Christianity, glorify violence, encourage drinking or smoking, flaunt elicit sex, and so on. A succinct letter to the network, local station, or government agency most appropriate to the situation can produce change. *Discovery!* newspaper (March/April 1990) reported that a boycott promoted by Christian Leaders for Responsible Television (CLEAR-TV) produced reduced amounts of sex, violence, and profanity in the Fall 1989 prime-time network broadcasts. The American Family Association (Tupelo, Mississippi) closely monitors network program content for such issues.

We can no longer afford to trust ungodly network executives and overburdened government agencies to police program content effectively. It simply won't happen. The standards of a secular society are much different than those of God's Word. It is time for Christians to begin consistently evaluating the messages that are broadcast. No longer is adult involvement optional. Increased monitoring and open discussion of the shows that we watch while reclining in our living room easy chairs is urgent and long overdue.

The addresses of your local stations should be listed in the telephone directory under the station's call letters.

Others are:

For children's shows on Saturday mornings:

- ABC Television
  Attn: Director of Children's Programming
  Century City, CA 90067
- NBC Television
  Attn: Director of Children's Programming
  Burbank, CA 91505
- CBS Television
  Attn: Director of Children's Programming
  Television City, CA 90036

For all other programming:

- ABC Television
  1330 Avenue of the Americas
  New York, NY 10019
- NBC Television
  30 Rockefeller Plaza
  New York, NY 10019
- CBS Television
  51 W. 52nd St.
  New York, NY 10019
- 20th Century Fox Broadcasting
  10201 W. Pico Blvd.
  Los Angeles, CA 90035
- Federal Communications Commission (FCC)
  1919 "M" St., NW
  Washington, D.C. 20590
- National Advertising Division of the
  Council for Better Business Bureaus
  845 Third Ave.
  New York, NY 10022

DISCUSSING PROGRAMS

While discussing TV programs may seem a little awkward at first, it is well worth the effort. The best thing that we can do to make better use of television is to watch what's on with discernment and discussion. In his book, *Taming The TV Habit*, Kevin Perrotta explains that, "We will be best able to make television viewing a positive and useful experience if we view it with other people. . . .We can develop habits of discussing what we are seeing, good and bad, in ways that heighten our understanding of the moral and other issues involved." Many harmful effects can be neutralized or lessened when viewers openly discuss and question TV content with friends and family members of all ages.

Frank discussions of TV programs often present many other opportunities for deeper conversations on a wide variety of topics that aren't likely to come up elsewhere. Fascinating exchanges on subjects like morality, corporate ethics, sportsmanship, marital infidelity, secular bias, consumerism, and attitudes become engrossing personal exchanges that can alter what goes into the minds of those sitting in front of the TV.

PROGRAM EVALUATIONS

Envision a youngster pleading to watch a show that you suspect will fall far short of God's expectations of decency or purity. You decide to watch and co-evaluate the program with your child. To emphasize the fact that **God** sets the standards (that this is not simply a clash between the "old-fashioned" opinions of parents and the "modern" opinions of a new generation) turn to God's Word before turning to the channel selector. Together, read from such passages as Ephesians 5:3-12 where you are reminded of the need for Christians to remain pure, both physically and mentally. Hand the child a Program Evaluation Form and instruct him to turn the show off if and when he feels that it violates scriptural principles. They'll enjoy the sense of responsibility and grow through the new sense of decision making.

Gregg Lewis, in his book *Telegarbage*, shares that, "...a junior-high son of one Christian father wanted to watch one of the more explicit adult sit-coms. His father hesitated, then agreed—if they watched it together and if the boy promised to keep a tally of every suggestive or shady line he noticed. Less than halfway into the show, the boy turned to his father and said, 'I see what you mean. I've counted fourteen already. Why don't we turn to something else?' They came to a joint agreement not to watch the program anymore."

While discussion certainly won't remove every negative influence of TV viewing, a combination of quality and quantity control can help to change the family TV from an ominous threat into a much more positive force. By taking time to co-evaluate programs, you exhibit loving concern. By reminding family members that God's Word (not yours) is the ultimate standard by which our everyday activities are to be judged, you reassure them that you are not simply intent on winning an argument. Rather, you are seeking to fulfill your own God-given responsibilities and to illustrate a practical application of God's Word.

USING THE "PROGRAM EVALUATION FORM"

If we are going to watch television, we must take time to determine the difference between the main plot and the subplots. When the messages are recognized, their power to influence us subconsciously is vastly diminished.

In a movie where the hero is involved in immoral activities, or where good is accomplished by deceptive actions (situation ethics), or crude jokes and profanity are used to make us laugh, we need to question seriously the wisdom of watching such actions. We need to remind younger viewers that the actions or language are not appropriate or acceptable in the sight of God, even though the writers are subliminally trying to tell us otherwise.

The purchaser of this book is encouraged to photocopy the following Program Evaluation Form for his/her personal use. Use an enlarging photocopier to increase the size so that it fits on an 8-1/2 x 11 inch piece of paper. Then, store copies in a 3-ring notebook to be kept near the TV. For the next several weeks, each time you turn on the TV, pick up an evaluation form. Then, analyze the content of the shows that you watch. In a very short time you will see beyond the obvious storyline. You will be amazed at the subplots, biases, ungodly actions, and values statements present in almost every form of secular media presentation. You will also develop a much greater degree of control over the television set. . .and it will gradually loosen its grip on your prime time.

Before shoving a pile of evaluation forms at your family, it would probably be wise to evaluate a few shows yourself. Read the sample form at the end of this chapter to get an idea of what type of data can go in each section. Get the hang of it before you try to interest others in such a project. If possible, try not to let anyone else know what you are doing.

Curiosity will soon get the best of not only your youngsters (if you have any), but also any other adults who happen to be watching with you. Once this happens, respond enthusiastically. If they express interest, ask if they would help you spot the items listed on the form.

Begin by outsmarting the network advertising executives. Use the commercial breaks as "time-out" periods. Turn off the TV or mute the volume and share with one another exactly what appears to be taking place on the program. Then, you might try to evaluate the commercials. Often, they are even more sexually suggestive and materialistic than the program you are watching. For offensive advertising, write the product sponsors. Tell them you find their ads tasteless or offensive and that contrary to inducing you to purchase the product, the nature of the ad has made you determined to avoid it altogether.

Above all, try to be casual about your observations and discussions, especially at first. Try not to intimidate or bombard other viewers with too many questions. Go easy unless you are absolutely convinced that a program is harmful or blatantly inappropriate.

Be forewarned. There is bound to be someone in almost every group who will complain that thoughtful discussion during dramatic presentations will destroy the enjoyment usually sought by viewers. Be sensitive. At first, you may need to agree to hold in-depth discussions for the commercial breaks or until the show is over. However, use of the Program Evaluation Form, because it encourages the exchange of ideas and opinions, actually begins to *enhance* the pleasure received by all viewers involved. A fuller understanding of all aspects of a program will allow much greater fulfillment, even if the show is not so good.

Finally, there is a side benefit to filling out this form. You will be *doing* something! Because you are actively evaluating the programs, you will tend to eat less junk food. Your mind remains active and alert. You no longer crave ice cream, potato chips, or brownies. Instead, you become engrossed in the project at hand. Don't be surprised if you lose a couple of pounds in the first week or so!

WHY NOT JUNK IT?

This is a very good and logical question. In fact, many viewers have followed through with this line of reasoning in a very literal way. A pastor friend of mine once confided that when he was a young boy his father took him, his brother, a sledge hammer and the family's only TV set into the back yard. The boys watched as their father—at wits' end due to TV contact—smashed the television into a shiny junkpile.

However, the long term solution to media-related problems does not seem to be that of banning TV altogether. While those few Christian families that have chosen to live without a television set are to be respected for their strong standards, they should also be warned of the need to sit down with their children and explain their reasons for not having one. After all, a complete, legalistic ban on TV may just be avoiding or postponing the inevitable problems instead of dealing with them. Children can usually visit friends to watch whatever they want. Many grow up, get married, and buy a TV for their own homes without ever developing discernment or good judgment on how to use or control it. (By the way, my friend's father eventually replaced the set that he had destroyed. In fact, the same father is now a retired grandpa with one "empty nest" and *five* television sets!)

Rather than junking it, the best overall solution seems to be that of training ourselves and our children to exercise biblical, moral judgment and quantity control. As TV reviewers, your family will develop several skills. They will learn to look carefully at the shows—and at other issues in life. They will begin to discern and discriminate and will have practice in putting their ideas into words as they start to realize that every statement, every hand gesture, every joke is intentional. Behind every line and every character is a meaning. Every moment of TV content has been designed by highly compensated scriptwriters and directors to communicate a specific message. Train yourself and your family to be alert enough to discern just what that message is—and the effect it's having on our society.

SAMPLE "PROGRAM EVALUATION FORM." Completed by the authors during and after viewing BABY BOOM.

PROGRAM EVALUATION FORM

THIS FORM IS DESIGNED TO AID IN EVALUATING THE CONTENT OF TV PROGRAMS,
VIDEOCASSETTES, THEATER MOVIES AND LIVE STAGE SHOWS.

BABY BOOM — _1-23_
Title of Show   Name of this Episode   Date Viewed

Dale - KM, MT, AJ, KMM
Evaluated and Viewed by

_7:30_ AM/PM   _9:20_ AM/PM   If Film _PG-13_   If TV _____
Time Began   Time Ended   or Video (Rating)   Show (Network)

SECTION ONE:
Complete this portion while show is in progress. Try to involve other viewers by asking for help in spotting the following items.

PROFANITY   Ephesians 4:29 & 5:4, James 5:12

1	2	3	4	5	6	7	8	9	10	11	12	13	14	15	16	17	18	19	20
✓	✓	✓	✓	✓	✓														

VIOLENCE   Psalm 101:3-4, Galatians 5:19-26

	1	2	3	4	5	6	7	8	9	10	11	12	13	14
FIST FIGHTS / KNIFINGS	∅													
GUNFIRE / EXPLOSIONS	∅													
RAPE / INCEST	∅													
MURDER / KILLING	∅													
OTHER VIOLENCE	∅													

Why are you still watching this lousy program?

SEXUALLY SUGGESTIVE CLOTHING (or Lingerie)   I Timothy 2:9

1	2	3	4	5	6	7	8	9	10	11	12	13	14	15	16	17	18	19	20
∅																			

NUDITY or SEXUAL INTERCOURSE (Implied or Explicit)   Ephesians 5:3

– TYPE OF RELATIONSHIP –	1	2	3	CONTEXT(s)
PREMARITAL (Fornication)	✓	✓	✓	_star w/ veterinarian, w/ boyfriend_
EXTRA-MARITAL (Adultery)				_in Condo, "nanny" w/ sailor_
ALTERNATIVE (Homosexuality)				
RAPE				
WITHIN MARRIAGE				

OCCULT PRACTICES (Magic Spells, Witchcraft, etc.)   Deuteronomy 18:10-14, Leviticus 20:6

1	2	3	4	5	6	7	8	9	10	11	12	13	14	15	16	17	18	19	20

OTHER ACTIONS OR STATEMENTS OF DISRESPECT TO GOD, OR TO CHRISTIANITY   Galatians 6:7, Isaiah 45:23b-24

① _Potential adoptive parents who refer to "pastor friend" (to allign them w/ Christians?) are backwards w/ dogmatic ways._
② _Shows potential nanny as fanatic Christian in a comical job interview scene._

SECTION TWO:

To be completed immediately after viewing, even if you are unable to complete Section One. Attempt to involve all viewers.

WHAT PROMPTED YOU TO VIEW THIS PARTICULAR SHOW?

___ Unplanned. Just bored and flipping through the channels.
___ This was a rented video tape. The label caught my attention as I browsed in the video store.
___ Saw advertisement in newspaper or during another show.
✓ Pressure from family or friends who wanted to see this show.
✓ Recommended by a friend.
✓ Other *Wanted to see a movie. This one was recommended.*

BASIC / SURFACE PLOT (most obvious storyline): *Yuppie female is forced to quit her 6 figure executive position after she becomes bonded to "inherited" niece. Leaves big city to live in the country. Meets man, starts successful business, decides to run her national enterprise from country home + be mother too.*

SUBPLOTS (the less obvious storylines or values statements): *① Sex before marriage is glorified. ② Living together w/ no intention of marriage is good if both partners agree. ③ Don't feel guilty about selfishness. ④ Christians are either legalistic, backwards, fanatical — or all three!*

OVERALL GENERAL IMPRESSION	Ephesians 5:15-17. Philippians 4:8		
		YES	NO
1. Did you seriously consider an alternative activity before viewing this show?			✓
2. Would Jesus, your parents, or your spouse approve of your viewing of this show?			✓
3. Was this show "pure, lovely, praiseworthy"?			✓
4. Do you now think that you should have more earnestly pursued an alternative activity?		✓	

MISCELLANEOUS COMMENTS: *We took both of our young children based on recommendation from good friends that they "have never seen a cleaner show". DECISION → Never take kids to a PG-13 film. There was no blatant nudity, but very effectively implied. Teaches Godless morals.*

OTHER SUGGESTIONS:

· COMPARE EVALUATIONS AND ANSWERS - Have each viewer complete a Program Evaluation Form as they watch the same show. Compare. It can be very interesting to see the same show, but to understand its messages differently.

· WHEN AT A THEATRE MOVIE - Try to sit far enough away from other audience members so that your writing doesn't become a distraction. Always complete Section Two as soon as possible upon completion of the show, even if there was not enough light to complete Section One while at the theatre. (Excellent discussion starter if you go out for refreshments afterwards!)

IMPORTANT! Trying to be careful about what they view, friends and relatives often ask for comments about theatre movies. By completing this form while the story is still fresh in your mind you will be able to give a much more fair endorsement or caution to those who ask.

PROGRAM EVALUATION FORM

THIS FORM IS DESIGNED TO AID IN EVALUATING THE CONTENT OF TV PROGRAMS, VIDEOCASSETTES, THEATER MOVIES AND LIVE STAGE SHOWS.

Title of Show _____   Name of this Episode _____   Date Viewed _____

Evaluated and Viewed by _____

_____ AM/PM   _____ AM/PM
Time Began   Time Ended

If Film or Video _____ (Rating)

If TV Show _____ (Network)

SECTION ONE:

Complete this portion while show is in progress. Try to involve other viewers by asking for help in spotting the following items.

PROFANITY
Ephesians 4:29 & 5:4. James 5:12

1	2	3	4	5	6	7	8	9	10	11	12	13	14	15	16	17	18	19	20

VIOLENCE
Psalm 101:3-4. Galatians 5:19-26

	1	2	3	4	5	6	7	8	9	10	11	12	13	14
FIST FIGHTS / KNIFINGS														
GUNFIRE / EXPLOSIONS														
RAPE / INCEST														
MURDER / KILLING														
OTHER VIOLENCE														

Why are you still watching this lousy program?

SEXUALLY SUGGESTIVE CLOTHING (or Lingerie)
I Timothy 2:9

1	2	3	4	5	6	7	8	9	10	11	12	13	14	15	16	17	18	19	20

NUDITY or SEXUAL INTERCOURSE (Implied or Explicit)
Ephesians 5:3

– TYPE OF RELATIONSHIP –	1	2	3	CONTEXT(s)
PREMARITAL (Fornication)				
EXTRA-MARITAL (Adultery)				
ALTERNATIVE (Homosexuality)				
RAPE				
WITHIN MARRIAGE				

OCCULT PRACTICES (Magic Spells, Witchcraft, etc.)
Deuteronomy 18:10-14. Leviticus 20:6

1	2	3	4	5	6	7	8	9	10	11	12	13	14	15	16	17	18	19	20

OTHER ACTIONS OR STATEMENTS OF DISRESPECT TO GOD, OR TO CHRISTIANITY
Galatians 6:7. Isaiah 45:23b-24

68

SECTION TWO:

To be completed immediately after viewing, even if you are unable to complete Section One. Attempt to involve all viewers.

WHAT PROMPTED YOU TO VIEW THIS PARTICULAR SHOW?

___ Unplanned. Just bored and flipping through the channels.
___ This was a rented video tape. The label caught my attention as I browsed in the video store.
___ Saw advertisement in newspaper or during another show.
___ Pressure from family or friends who wanted to see this show.
___ Recommended by a friend.
___ Other _____

BASIC / SURFACE PLOT (most obvious storyline): _____
_____
_____
_____
_____

SUBPLOTS (the less obvious storylines or values statements): _____
_____
_____
_____
_____

OVERALL GENERAL IMPRESSION	Ephesians 5:15-17, Philippians 4:8		
		YES	NO
1. Did you seriously consider an alternative activity before viewing this show?			
2. Would Jesus, your parents, or your spouse approve of your viewing of this show?			
3. Was this show "pure, lovely, praiseworthy"?			
4. Do you now think that you should have more earnestly pursued an alternative activity?			

MISCELLANEOUS COMMENTS: _____
_____
_____
_____
_____

OTHER SUGGESTIONS:

• *COMPARE EVALUATIONS AND ANSWERS - Have each viewer complete a Program Evaluation Form as they watch the same show. Compare. It can be very interesting to see the same show, but to understand its messages differently.*

• *WHEN AT A THEATRE MOVIE - Try to sit far enough away from other audience members so that your writing doesn't become a distraction. Always complete Section Two as soon as possible upon completion of the show, even if there was not enough light to complete Section One while at the theatre. (Excellent discussion starter if you go out for refreshments afterwards!)*

IMPORTANT! *Trying to be careful about what they view, friends and relatives often ask for comments about theatre movies. By completing this form while the story is still fresh in your mind you will be able to give a much more fair endorsement or caution to those who ask.*

CHAPTER SEVEN:

HOW MUCH DO YOU REALLY WATCH?

Ask a mountaineer why he feels compelled to scale a mountain and, "Because it's there!" comes the response. Many TV owners turn the tube on for the same reason.

For a very high percentage of TV viewers, habit dictates action. For many people, TV controls them more than they control it. Do you or some of the members of your family fit into this category?

SIGNS OF ADDICTION

Honestly answer this self test to check for over-use and addictive tendencies. A "yes" response to any one of the following questions should serve as a warning flag that improvement is needed. Answering "yes" to at least three questions strongly indicates that you should seriously re-evaluate your viewing habits.

QUESTION   YES NO

1) Do I frequently turn on the television without knowing what's on, flipping through the channels looking for the most appealing show that happens to be on at the moment?   __ __

2) Do I sometimes fail to turn off a program which is uninteresting or offensive? ___ ___

3) Do I usually leave the set on when someone comes to visit? ___ ___

4) Do I keep the TV on during meals with others? ___ ___

Many women are surprised after answering these questions. It is not only men and children who use TV improperly. Female viewers often need to restructure their viewing habits just as urgently as the rest of the family. In fact, 41% of the readers of *Glamour* magazine, almost all of whom are women, consider themselves to be "couch potatoes." In answering a February 1988 *Glamour* survey, the women admitted, at least when polled anonymously, that they aren't exactly positive role-models in this area.

And yet, positive role models are exactly what the younger members of our households so desperately need. Children are profoundly affected by their parents. If Mom and Dad tend to watch television extensively, the children will follow that model.

WHAT ARE YOU MISSING?

In one *TV Guide* study, researchers mounted cameras on top of TV sets and recorded the amount of time viewers also read, talked, walked in and out of the room, and so forth. They concluded that viewers actually *watch* only 55% of what is on.[1] Television jabber often serves as a background during other activities such as doing homework, housework, cooking, or balancing the checkbook. In most homes, having the set on becomes a ritual, a habit.

The sheer amount of time spent with the TV on is crucial not only because of the potential harm, but also because of the many good possibilities missed. For most of us, "our heavy investment in television viewing squanders our time. It is a bad deal. What we gain in entertainment and information does not compensate us for what we lose in terms of ordinary life."[2]

When the TV is on, our reading is distracted and superficial. We usually don't talk with family and friends or in depth about subjects of great importance. We don't fix the roof or play softball with the kids outdoors. We don't spend quality time thinking or planning. We don't gather for family devotions. We don't benefit from quiet times with our Creator. First Corinthians 6:12 challenges us, "'Everything is permissible for me'—but not everything is beneficial. 'Everything is permissible for me'—but I will not be mastered by anything." This chapter is designed for a personal, honest evaluation of what or who has mastery in our lives.

Regardless of who we are, we all have duties to perform, people to care for, and our God to know better. It stands to reason, then, that if our lives and responsibilities are important, the time God gives us on this earth to live and carry out our responsibilities must be important, too! We should ask ourselves frequently if spending time with the TV might be costing us more than we should be willing to spend. We need to pursue the things that God might want us to do, rather than constantly pursuing the things that we want for ourselves.

KEEPING A "VIEWING INVENTORY"

We need an accurate inventory of the amount of TV viewing done by each person within the household. Parents usually underestimate the number of hours that they and their family spend in front of the set. Those adults who keep a running record for a week or a month are generally surprised at how the hours add up.

There are times when we need to just stop and do nothing. And television viewing certainly fills that need well. However, the question, as one eloquent author put it, is: "Do we need to do nothing for several hours every day?"

Generally speaking, the following table defines your skills in TV management, *assuming that you are reasonably healthy* **and that the only shows viewed are of a wholesome, scripturally acceptable nature.** (Filling out a "Program Evaluation Form" from the previous chapter will help you more accurately determine what is scripturally acceptable and what is not.) Additionally, although viewers who watch 0-7 hours are rated as "Excellent" TV managers, this in no way means that one must meet some sort of quota every day. If you find yourself feeling compelled to watch TV daily—even for just one hour—you may be controlled by it.

HOW DO *YOU* RATE?

Hours spent per person per week with TV on	Type of Viewer
0 - 7	*EXCELLENT* TV Manager
8 - 14	*ACCEPTABLE* Moderate cutbacks helpful
15 - 21	*HEAVY USER* Serious cutbacks needed
22 or more	*ADDICT!!* **Terrible TV Manager** Urgent restructuring necessary

The buyer of this book is granted permission to photocopy the following chart for the limited use of his/her family. By using an enlarging photocopier and then posting the full-size (8-1/2 x 11) "Viewing Inventory" forms in some prominent place (on the refrigerator, on the living room coffee table, taped to the TV set, etc.), you can keep track of every minute that your TV is on. Do this for at least one full week, preferably three or four. Heavy viewers will have to write small or use more than one chart to record honestly everything that they watch.

The simple act of consistently filling out this form encourages pre-selection of programs over channel scanning. It also helps to cut down on the amount of time the TV is on but not really watched.

After reading the next chapter and initiating some of the suggestions listed there, complete another chart. You will be pleasantly surprised at how quickly and dramatically your viewing habits improve.

FOOTNOTES:
[1]*TV Guide* (May 30, 1987); p. 5.
[2]Kevin Perrotta, *Taming The TV Habit*, (Ann Arbor, Michigan: Servant Publications, 1982), p. 39.

VIEWING INVENTORY

A DAILY LOG OF HOURS SPENT WITH THE TELEVISION ON

DATE		NAME OF TV SHOW (or video)	TIME BEGAN	TIME ENDED	TOTAL
	MONDAY				
	TUESDAY				
	WEDNESDAY				
	THURSDAY				
	FRIDAY				
	SATURDAY				
	SUNDAY				
		TOTAL NUMBER OF HOURS THAT THE TV WAS ON DURING THIS ONE-WEEK PERIOD			

HOW DO YOU RATE?

Hours spent per person per week with TV on	Type of Viewer ★
0 - 7	**EXCELLENT** TV Manager
8 - 14	**ACCEPTABLE,** Moderate cutbacks helpful
15 - 21	**HEAVY USER,** Serious cutbacks needed
22 or more	**ADDICT! / Terrible TV Manager** Urgent restructuring necessary!

★ Assuming that all programs viewed are of a scripturally acceptable nature. (Refer to Chapter 3 for biblical guidelines.)

Your Name _____

CHAPTER EIGHT

WITHDRAWAL SYMPTOMS SURVIVAL GUIDE

If you have decided that now is the time to begin restructuring your personal or family television viewing habits, congratulations! You are about to join the ranks of a select group.

Only a small percentage of the tens of millions who need to change are actually willing to admit the need, even to themselves. But for the many who have learned how to better control their use of TV, they consider the rewards—relational, emotional, spiritual, and financial—immeasurably worth the struggle it entailed. Take it from one who's been there, though, it isn't easy, especially at the beginning of your "TV diet."

For most of us, television is deeply entrenched into our lives. This didn't happen overnight. It took years. Our habits regarding its use are not easily modified. In fact, Dr. Pierre Mornell, a California psychiatrist and knowledgeable researcher on this subject, says that if you are used to having the TV on several hours a day, "getting rid of it cold turkey is likely to cause a seizure."[1] While his overstatement makes us chuckle, almost everyone who has seriously undertaken to limit his viewing to an average of 1 to 1-1/2 hours a day has had to cope with the symptoms of withdrawal.

How long these symptoms last and how severe they will be is usually unpredictable. They attack the entire system, often showing up as irritability, extreme nervousness, moroseness, depression, apathy, appetite loss, excessive withdrawal into sleep, and others. While most adults can imagine these types of symptoms in children, they are not limited to youngsters. The Society for Rational Psychology in Germany paid volunteers to abstain from television for a year, and discovered that adults also suffer withdrawal.[2] But be encouraged. The rewards are well worth the sacrifice! To help you on your way, this chapter contains ten practical strategies for winning your struggle.

PREPARING FOR VICTORY

Satan does not want you to be selective in your use of television. Remember, he's the prince of the power of the air—and, today, all too often the air waves. He has worked hard to develop a generation of remote control-clutching men, women, and children. Our adversary won't give up without a struggle. He knows that his years of striving to "hook" us on afternoon soaps, evening sit-coms, weekend sportscasts, and violent- or occult-oriented children's cartoons will be a failure if you choose to take the steps necessary to restructure your prime time.

It may sound like an overstatement, but *you must begin by resolving in prayer before the Lord (preferably as an entire household) that you will strive to watch only those programs that measure up to the biblical directives set forth in God's Word.* Then, ask someone you can depend upon— your spouse, your parents, a close friend, Sunday School teacher, pastor, even your children—to pray for you. Next, give them a copy of this book. Ask that they read it and that they help you by holding you accountable, checking periodically to see how well you're doing.

If you are married and your spouse has not yet come to the same determination to restructure his television viewing habits, politely, and in love, ask if he/she, too, would read this book. By doing so, your spouse will much more fully understand why you are doing what you are, and how she can best help you.

MY LIFE HAS CHANGED

My own attempt to restructure a deeply imbedded TV habit initially met with frequent frustration. It took time, prayer, and a lot of searching for help in my struggle. Again and again my good intentions failed and confirmed the well-known adage that "old habits die hard." I might go several days, even a week or more, but soon a night would come when I would give in to the same old temptation.

Following Karen to our bedroom, I would kiss her goodnight and assure her that "I'll be in in just a few minutes." Then, after creating a generous Dagwood sandwich, I would find myself drawn into the same selection of late night viewing options as millions of others. "I'll just keep the set on until I finish my midnight snack" was my usual logic. However, when the snack was gone, all-too-often I would lay a toss-pillow on the coffee table and prop my big feet where I knew they shouldn't stay.

Redirecting my desire to watch TV proved to be extremely difficult. *Everything* else seemed like work when compared to the effortlessness of sitting down with the remote control unit. I discovered that I had gotten so used to triggering the "on" button as soon as I came through the door at the end of each work day that I had all but forgotten the myriad of other, more rewarding things to do. The problem was that the last thing that I was likely to do when

I felt like watching TV was to get creative and try to think of something else. None of the books I read had prepared me for the withdrawal symptoms that habitual TV watchers almost always experience when we begin to realign our evening and weekend hours.

But slowly—at first reluctantly—I began to insert other activities where TV once was. Rather than turning my attention to the evening news as soon as I got home, God led me to a series of Christian novels that I began to read to Karen as she finished dinner preparations. Rather than running for the *TV Guide* as soon as the last morsel had been consumed from my dinner plate, I carried my plate to the kitchen. *(I discovered that wives can give great hugs if husbands help clean up the supper dishes!)* Then, rather than turning up the volume to quiet (or drown out) a bored baby, I found that unfolding the collapsible stroller and strapping a diaper-clad damsel in for a family walk usually calmed even the most frantic cries for attention. And those evening strolls eventually led me to a determination to exercise more frequently.

In effect, what I learned was that simply telling myself that my use of the TV should be restructured was not enough of an incentive to keep me true to my goal. Without other, *appealing* activities, I inevitably fell back into the same old rut. I had leaned on TV to fill my "free time" for far too long, so I needed ideas for entertaining alternatives and reminders of responsibilities. Karen could sense this need and began to compile a list of "TV Alternatives" that we kept handy and could easily scan whenever we felt the urge to turn to the tube. Section IV of this book lists some of her ideas. We're sure you'll have more of your own.

We still watch TV, but it has become a tool rather than a crutch. We learned how to turn it off—and on only when a show that we especially wanted to see was scheduled to be aired. We continue to watch videos, but we have become much more selective about program content and messages.

WITHDRAWAL SYMPTOMS SURVIVAL METHODS
Because we know that the pull of the TV can be very intense, we strongly suggest that you initiate at least two or three of the following methods for surviving the initial symptoms of TV withdrawal. All of the ideas are viable, and most will cost you little or no money to institute. Choose a couple that best suit your needs.

Method #1 - ASSEMBLE A "SURVIVAL KIT"
This is the most basic and probably one of the most practical ways to help insure victory in your struggle to adjust your use of television. Simply grab a cardboard box, milk crate, or even a large, unused tool box and fill it with as many of the following items as you have on hand. Organize it as neatly as possible, and then place it in the TV-room. Its very presence suggests that something other than TV be turned to throughout the day. Make sure that it is easily accessible, and be sure to commend those who use it.

SURVIVAL KIT - Suggested Contents:
1) COLORING BOOKS
2) CRAYONS, colored pencils, or markers
3) 3-RING NOTEBOOK with:
   - blank "Program Evaluation" forms
   - blank "Addiction Inventory" forms
   - a "fun to-do" list with suggestions of activities that you know various members of your family will enjoy
   - a "chores to-do" list, reminding each child of required household duties

- names and phone numbers of people whom you would like to get together with for non-TV activities
4) THIS BOOK, with the "TV Alternatives" section clearly marked
5) CRAFT PROJECTS
   - model cars, airplanes, ships, rockets, etc.
   - needlework, mending needs, craft kits
6) CRAFT AND HANDYMAN MAGAZINES
7) STATIONERY (or other writing paper)
8) ENVELOPES, with stamps already attached
9) TAPE RECORDER
10) BLANK CASSETTES (on which to compose "audio letters")
11) RECORDED AUDIO CASSETTES
   - Christian music
   - dramatized stories
   - Bible on tape
   - sermons
   - favorite speakers
12) BIBLE
13) BIBLE STUDY WORKBOOK(s)
14) CHRISTIAN NOVELS (C.S. Lewis, Janette Oke, David and Karen Mains, B.J. Hoff, Bodie Thoene, etc.)
15) CHILDREN'S STORY BOOKS
16) RADIO (pre-tuned to a Christian or other "parent approved" station)
17) JIGSAW PUZZLES (for various skill levels)
18) BOARDGAMES
19) RAFFLE TICKETS (to be used as described in "Method #4" of this chapter)

Method #2 - TWO-WEEK SHUTDOWN
It is often much easier to manage the fingertip availabil-

ity of TV if it is completely out of reach for a reasonable period of time at first. Physically unplug your set(s), turn them toward the wall, even remove them to a closet, outdoor storage shed, or a neighbor's garage.

If you have children, be sure to have a family conference during which you explain why you are taking what will undoubtedly appear as a drastic step. Admit your own misuse (if applicable), and commit publicly to your goal of being a much better example in the future.

When the TV is re-introduced into the household, limiting the amount of time it is on and determining what programs are allowed to be viewed will be an easier task. Going from no TV to some TV is usually much better than cutting back from limitless viewing to your totally new standards. Also, the family will know that it is actually possible to live without this modern technology and will probably try some new or forgotten activities.

Method #3 - TV VIEWING CONTRACT

I recently read about a grade school principal in the Midwest who orchestrated an amazingly successful week of TV abstinence in her school of 270 students. Mrs. Pat Sumrow of Edison School in Elmhurst, Illinois used a simple contract, the "Black Out Television Pledge Card," that was sent home with her students. Two hundred forty-nine of her 270 students committed to stop watching TV for one week by signing their names to the contract and by getting one of their parents to sign, too. An adult's signature space was included on the form so that teachers were assured that someone from home would help the young people stay true to their goal. The incentive for the students: a sleep-over for all successful abstainers, chaperoned by PTA members and teachers. Over 99%, 247 of the 249, actually made it through the week "TV free"!

You may want to try to interest your children's teacher or principal in a similar challenge. Or, you can design a scaled down version in which you try to interest your child's primary playmates. Type up your own contract. Offer to take your child and any of their friends on a weekend campout, water skiing, or some other activity, if they can go for one or two weeks without turning on the TV for any reason (including videos and electronic games).

The positive peer pressure generated by a challenge such as this is very beneficial, especially if you initiate the "Two Week Shutdown" mentioned previously. This contract idea is particularly well-suited to older children, since Junior High and High School youths may balk at the following method.

Method #4 - TWENTY TICKETS

Dr. James Dobson, widely respected as one of our nation's foremost advocates for the family, has suggested a system for controlling television. In his organization's *Focus on the Family* magazine, he has repeatedly endorsed a very practical, easy to initiate program to: 1) monitor the *quality* of programs that our children watch, 2) regulate the *quantity* of television they see and, 3) include the entire family in establishing a TV policy.

In essence, his "system" is this: First, adults sit down with their children and select a list of parent-approved programs and videos that are appropriate for each age level in the home. Use the weekly *TV Guide* or your local newspaper's TV schedule to determine this list.

Then, highlight the approved shows, type, or write them on a piece of paper and place your list on the refrigerator or somewhere else it can be easily referred to throughout the week.

Third, buy a roll of inexpensive raffle tickets (usually

available from any business supply or party supply store) or make a stack of your own by using 3 x 5 cards. Issue 20 tickets to each member of your family (including Mom and Dad!) each week for at least a month. Each ticket can be used to buy up to 30 minutes of program viewing from the approved list. When a child has "spent" his 20 tickets, he has completed his television viewing for that week. This system reduces parental "nagging" and forces each individual to be selective, carefully choosing what programs he most wants to spend his time watching. Of course, neither Mom or Dad can give in to the inevitable whining if the tickets are spent too quickly.

Dr. James Dobson of Focus on the Family feels that ten hours a week is probably a good target amount, but encourages parents not be too rigid or legalistic in their approach.[3] A special program or a holiday program during Christmas or Thanksgiving, may mean more tickets or counting it as "free viewing." You might also give extra tickets as rewards for achievement. At our house, we don't make anyone "pay" with a ticket when they exercise to an aerobics video.

An added inducement for children not to use all of the tickets would be for you to "pay" them something for each ticket that is returned *unused*. The rewards could range from a certain amount of money for each ticket, to a special dessert one evening, to the privilege of an extra-special family outing (roller skating, bowling, amusement park, camping, visit to a museum, etc.) for the accumulation of a certain number of unused tickets over a three or four week period.

Method #5 - THE EXERCISE CONNECTION
Many families already have one. If yours doesn't, just visit a few rummage sales next weekend. It's an exercise

bicycle. Slightly modified and connected to your television along with an automobile alternator and a 12-volt battery, you can pedal up some watts right in your own living room!

This has got to be one of the most unique methods yet devised to improve the way that families watch TV. It is guaranteed both to encourage selective viewing and to improve the viewer's physical fitness since every minute that the set is on must be "earned" by riding the bike to store energy in the auto battery. The TV then draws its electric current from the battery.

A set of plans that explains the set-up in detail can be acquired free of charge simply by visiting your local library. Go to the Reference Desk and ask for the microfiche for the March/April 1981 issue of *Mother Earth News*. Locate page 134 and you will find an article entitled "Cycle Power, Part II." In addition to an electrical schematic, a picture of what the connections should look like, and a detailed parts list, the article even describes how to build an exercise bike from scratch. It's a great way to "exercise" self-control for your family's TV viewing habits. It is also a super science fair project for any resourceful child or teen who enjoys a challenge!

Method #6 - INCONVENIENCE YOURSELF

If your TV is a lightweight or portable model, consider clearing a semi-permanent place for it on a laundry room shelf or in some other inconvenient location. Remove it from this spot only when there is a program on that is worthy of your attention. Return it to the storage place immediately after the show is over.

If the TV is too large or heavy to move, unplug the set, cover it, and use it as a table for a nice flower arrangement. This way you are forced to think twice before watching whatever happens to be on at the moment. If an "on" switch is not immediately available every time you enter the room, you will be less inclined to simply sit down and start watching.

Method #7 - THE SWITCH

Like many concerned mothers throughout our nation, Addie Jurs was frustrated by the way her three children allowed TV to consume their time and inject undesirable images and values into their pliable minds. While she enjoyed it for the educational and wholesome entertainment value that it could bring, to a large degree the TV set had become an unwelcome intruder in the Jurs household. Sound familiar?

Mrs. Jurs tried several ideas to diminish TV's strong influence over her children. Finally, Jerry Jurs thought of a sure-fire way that he and his wife could control when and how frequently the TV was used. Jerry bought a metal electrical box from the neighborhood hardware store, cut off the plug from the TV cord and "wired the cord into the box, which contained a key lock that turned the power off and on. The finished box didn't look like much, but it helped the family establish a reasonable TV policy."[4] Amazingly, the Jurs have discovered that their boys now sometimes *ask* them to lock the TV off when they have important schoolwork to complete.

Thrilled by their success, Mrs. Jurs began searching for a way to develop and market her husband's idea. She soon discovered that a California man who had gone through a similar experience had put into production a very nice looking, inexpensive key-lock device that turns the power to the TV on and off. It is installed by simply plugging the power cord into a child-proof compartment. Special screws make this device ("The Switch") almost impossible for the average child to unplug without mom and dad's knowledge.

"The Switch" is especially popular among parents of teenagers. It is also a convenient way to assure working parents that their children will not idle hours away in front of the TV while mom and dad are still at work.

If "The Switch" sounds like a good idea for your family, you can contact Addie Jurs at the address listed at the end of this chapter.[5] This unique device can also be used to lock-out computers, video games, or stereo equipment.

Method #8 - BARE "ESSENTIALS"

Unless you have an invalid in your home, or some other very unique situation, you probably do not need to have more than one television. Yet, an amazingly high percentage of Americans clutter the kitchen counter, the bedroom dresser, the living room corner, even the garage workbench with a wide variety of TV sets. Some have stereo speakers in oak cabinets, some are nothing more than cheap neon-colored plastic, but all are capable of exerting almost inescapable magnetism toward those around them.

People have varying reasons for having multiple sets in the home. We have heard families say that additional TV sets are more convenient. Family members no longer fight over which program to watch. But the bottom line is: "Do these extra TVs strengthen our fragile lines of communication, or do they further separate us?" "Do they tempt us to watch more TV than we ought?" "Is any amount of convenience that they bring worth the price of the relationships in our home?"

By selling off all of your extra televisions and video players, the programs that enter your home can be regulated much more easily. Also, the money that you receive can be used to pay bills, be donated to needy families, used to take the family on vacation, given to a ministry that is involved in strengthening the family, given to your local

87

church, or reinvested in board games, books, wholesome videos, uplifting music tapes, and so on.

When you stop to think about it, having a TV set in each main room of your home is about as foolish as a dieter stashing chocolate bars throughout the house. You are setting yourself up for an unnecessarily tough battle until the number of sets sprinkled throughout your home can be counted on a single finger.

Method #9 - ACCOUNTABILITY

As mentioned earlier in this chapter, the act of making yourself accountable to a friend or family member can be very helpful. Let someone else know that you are setting out to change the way that you watch TV. Ask them to hold you accountable by:

* checking weekly to see if you are keeping track of the number of hours and types of programs you view.
* seeing that you are discussing with your children/spouse/ friend the shows that they are watching to be sure you and they understand all messages—hidden and other- wise.
* talking about new activities you have tried during the week.
* sharing what you have learned in your Bible study that week.

You may even want to ask your friend to join you in your endeavor to improve TV-related habits.

Method #10 - KEEP THIS BOOK HANDY

This book has been designed to provide you with the essential tools (forms, TV Alternatives, Christian video evaluations) for your ongoing struggle to restructure your TV viewing habits. Don't file it away on a bookshelf once your initial reading is finished. Keep it handy.

FOOTNOTES:
[1]Nina Coombs, "Is Your Love Life Going Down 'The Tube'?" *Reader's Digest* (October 1987).
[2]Gregg Lewis, *Telegarbage* (Nashville: Thomas Nelson Publishers, 1977) p. 102.
[3]"Dr. Dobson Answers Your Questions" *Focus On The Family Magazine* (November 1988): p. 6, also (March 1987): pp. 6,7.
[4]"News and Such," *Focus On The Family Magazine* (July 1989): p. 11.
[5]For more information about how you can order "The Switch," contact: A.J. MARKETING
                267 Walker Ave.
                Clarendon Hills, IL 60514
                (708)323-0625

CHAPTER NINE

VCR: *FRIEND OR FOE?*

More than four decades have elapsed since TV ownership became affordable for the average American. During that time, corporate research and development divisions have been very busy. And very productive.

The electronic technologies surrounding TV and its related gadgetry have improved dramatically. Perfectly clear, full-color pictures have replaced fuzzy black and white haze. Digitized stereo sound has supplanted the crackling of the original in-cabinet speakers. Cable TV wires carrying anywhere from ten to one hundred or more additional channels now silently snake their way through our neighborhoods and into the living rooms of over 42 million American households. And satellite dishes, originally most popular in rural and mountainous areas where normal in-air reception is poor, now dot the rooftops of an ever increasing number of urban residences.

Add to these marvels of human ingenuity the tremendously popular video cassette recorder. These compact TV-top boxes are now wired between antenna and television set in approximately 75% of American households. "This fact is disheartening as well as encouraging. On the one hand Americans have greater access to sexually explicit and violent films. But on the other hand, . . .with the VCR, families have the capability of controlling and increasing the selection of entertainment they watch on TV."[1] If a Christian is going to have a television in the home, one of the best and most convenient alternatives to normal network fare is the videocassette player. However, because the VCR can be programmed with violence and raw sex just as easily as it can be programmed with uplifting materials, the ability to discriminate increases in importance.

VIDEO SEDUCTION

Millions of red-blooded Americans who would never slink through the front door of a smoke-filled "adult bookstore" or a triple X-rated film house can now tote the *same* graphic material home to the privacy of their own living room. And they do so without the embarrassment of being seen in the "wrong place."

One pastor, who chose to remain anonymous, wrote an extremely personal and revealing article about his own video seduction which originally appeared in *Moody Monthly* magazine. Finding himself the beneficiary of a VCR intended to help enhance his ministry, he took the machine home and hooked it up in the family room. He envisioned himself viewing messages by some of the most respected Christian leaders and purchasing some animated Bible stories for his children. He also wanted to "time shift" some of the better quality childrens' shows and family-oriented movies to an hour more in line with his family's schedule.

Driving to the nearest video rental store, he walked through the front door intending simply to pick up a Disney classic. However, as anyone who has ever been inside a video store can attest to, his attention was diverted by the provocative posters and explicit video jackets used to draw attention to all varieties of ungodly productions. At first he stood firm, but after a while, he began to note that many of his own congregation members were discussing the same films that he had been struggling to avoid. Eventually he began taking PG and PG-13 dramas home for him and his wife to view after the children had gone to bed. There was profanity and occasional nudity, but the two of them felt firmly rooted in their Christian walk so that such things wouldn't really have an affect on them. Or would they?

After a time, the once objectionable content of the films that they had been renting became commonplace. What was originally felt to be extremely violent or provocative was now only mildly so. One day he finally rented an R-rated move. Before long he found the sexual scenes more enticing and less offensive.

This pastor continued his daily personal devotions, but knew that it had really become a sham. His addiction and his cravings became more intense.

He was finally jolted into changing his habit and asking his wife's forgiveness after a weekend when she had gone out-of-town for a women's conference. The author of this true-life account writes that:

...I had not dared to bring an X-rated film home for us to view together, but now was my chance to view one alone.

I contemplated this decision for at least a week. I did not want to give in to this temptation. Yet I could not seem to get the film out of my mind.

On Saturday night I went to the video shop intending to get a family movie. But as I laid my three dollars on the counter for an X-rated movie, I told myself, "It's just curiosity, not lust. Perhaps as a Christian leader I should be aware of what the world is consuming."

What I saw was ugly. The film degraded men and women. The beauty of human sexuality as God designed it and as I had experienced it in marriage was absent. I felt empty, cheated, and defeated.

It was at this point that God brought me to my senses. He

had been calling me to repentance all along, but I had been ignoring Him. Shocked by my failure, I realized I was in danger of destroying my life and my ministry. If I hardened my heart and harbored this sin, what would entice me next?

I did four things that night before going to bed. First, I destroyed the identification cards that video shops require when renting tapes. Second, I wrote a letter to my wife, confessing my failure and asking her to pray for my spiritual recovery. Third, I confessed my sin to the Lord and appropriated His cleansing. Fourth, I made a decision before God to stay out of video shops.

I'm writing this not to provide a catharsis for my soul, for I'm assured of God's forgiveness (I John 1:9), but to warn other Christians who own video equipment. *No Christian is immune to the temptations of video seduction.*[2]

CHILDREN AT RISK

"A study by University of Maryland professor Mark Levy shows that almost all the prerecorded tapes watched by 10th graders are R-rated films."[3] Practically speaking, children are free to rent just about anything that they are tall enough to reach. The vast majority of secular video store clerks do not respect or even try to enforce the rating system set up in the late 1960s by the Motion Picture Association of America. To do so undermines the primary reason for their existence—profit! Once again, it is the responsibility of parents to train and monitor their children. The entrepreneur squeaking a living out of his neighborhood video rental business is not going to do it for us!

SO WHY OWN A VCR?

Kids love videos! An adult usually tires of a show after seeing it two or three times, but children will watch a video dozens of times.

Today's children run around the house hugging videos

the way their parents used to hug their favorite dolls or teddy bears. The VCR is definitely reshaping the way that children of all ages watch television. Recent statistics reveal that, in the United States, "VCRs are used an average of seven hours a week, with half of the time used to play rented or purchased tapes. Children spend 50% more time watching VCRs than their parents."[4]

The captivating nature of video has been expertly manipulated to teach humanistic values through a wide variety of shows which, at first glance, appear to be of a generally wholesome nature.

However, a wonderful and growing selection of high quality, Christ-centered and Bible-based videos are now available. The entire final section of this book has been devoted to evaluating over 100 of the best and/or most widely marketed Christian videos. No longer is it necessary for your family to be brainwashed with spiritually harmful messages when all they want is a short period of visual entertainment.

Parents, especially, can sense a feeling of security in knowing that there *are* programs available that they don't have to be on guard against. They can experience the thrill of allowing their children to sit down and watch dramas and cartoons that are carefully written to communicate spiritual truths through entertainment. With careful monitoring and limited but repetitive exposure to these programs, parents can now enjoy a refreshing sense of victory in their attempts to wean children away from their willful addiction to violence and magic/occult-oriented "children's cartoons."

Seeing Christian principles and biblical stories acted out on the TV screen can go a long way in helping to impress them in young, pliable minds. Additionally, seeing an animated or dramatized version of a well-researched Bible story gives it an even greater feeling of reality. And how refreshing to finally view people on your television screen that actually accept the presupposition of a God who is in control! What a wonderful reassurance to your children that their parents are not "weird" and that yours is not the only Christian family on the face of the earth.

A FEW ADDITIONAL TIPS

Call around to the Christian bookstores in your area to find out which ones rent family-oriented and Christian dramas on videocassette. By getting our videos from Christian bookstores, we can avoid most of the unnecessary temptations that were so well revealed by the author of the *Moody Monthly* article referred to earlier in this chapter. Also, the more that we rent from these family-oriented sources, the more that they will be encouraged and financially able to invest in additional titles to broaden their selection.

If you do own a VCR, become familiar with its operation so that you can quickly set the automatic timer to record specials and family oriented programs that are scheduled for broadcast at an hour which is otherwise inconvenient for you. This "time shifting" is vastly under-utilized by most VCR owners. Never again should you feel that you have to miss a church service or eat dinner in front of the TV set because a show that you simply "must see" is going to be aired a the same time.

Videos can be a very positive alternative to normal network and cable TV fare. But viewing even the best videos should not be seen as an alternative to family or personal times of other, non-TV activities. When carefully used, the videocassette player can be a tremendous supplement for family and spiritual enrichment.

Once a person decides to try to at least partially close the door to negative influences by initiating new TV viewing standards, he will find himself desperately searching for viable alternatives. Without other enjoyable activities in which to invest your time, your good intentions may be only that. It is to meet this very real need that Karen has designed the following "TV Alternatives" section.

FOOTNOTES:

[1]Bob Maddux, *Fantasy Explosion*. (Ventura, California: Regal Books, 1986), p. 138.

[2]Anonymous, "Video Seduction," *Moody Monthly* (May 1987): pp. 28-30. Used with permission.

[3]Dr. William Fore, "Videocassettes present a Challenge to the Church," *Christian Film & Video* (a newsletter of the Wheaton Graduate School of Communications), (Vol. 6, No. 1, January-February 1989): p. 2.

[4]Fore, p. 1.

CHAPTER 10

CHANGING OLD HABITS

Dale and I know from experience that old habits are hard to change. As of this writing, we are determined to shed a few pounds by changing some eating habits. I do want to lose some weight, be healthier, look better, but I still find myself magnetically drawn into the kitchen. There I stand scouring the refrigerator shelves in search of some forbidden snack. I am about to succumb to a culinary evil until my eyes fall upon a little bowl of carrot sticks. They wait humbly on the shelf, carefully prepared for just such a moment of temptation. I pick up the healthy alternative and enjoy one more small victory over an old habit.

Changing our television consumption is, in some ways, like changing old eating habits. We know that too much TV isn't good for us. We want to develop better communication and new abilities. We are sure that we want our children to do more than beg for junk food TV. But we are easy targets for failure unless we plan ahead for those moments of temptation.

This section contains a list of 150 ideas of things to do other than watch television. It is your resource for those moments of weakness. Use it to re-train yourself and your family in the arts of recreation, work, communication, togetherness, crafts, and service to others. Although some of the ideas are novel, many suggest ordinary activities of which we all need to be reminded.

It was important for our family to actually make a list of TV alternatives for those situations when it seemed impossible to recall even the simplest ideas. Like most mothers, I, too, have those days when creativity is far from me. The baby is clinging to my leg screaming at the top of her little lungs as I drag her around the kitchen desperately trying to work on dinner preparations. As I glance over my shoulder, I see my two-year-old finger painting in the flour she has just spilled all over the counter. Her dusty white face is intent on expanding this powdery disaster. I know that my husband will be walking through the front door any minute with that wearied look that says, "I'm going to die of starvation in five minutes! What's for dinner?!" And then my eldest whines, "Can we watch TV, Mommy?" At times like this, a strange thing happens. My mind becomes a blank fog. Except for the word "television," nothing else comes through clearly on my mental screen.

"How can it be," I wonder, "that a college graduate, trained in elementary education, cannot think of one activity for her children besides watching TV?" This uncanny loss of memory is what motivated me to compile a list of "TV Alternatives" as an easy reference and planning guide for moments of temptation. . .and desperation. And, I know that if you want to remember specific points, it helps to begin each one with the same letter. So here are six guidelines that I guarantee will change a lifetime of TV habits: 1. Pray for God's Help
2. Plunge In
3. Personalize This Book
4. Place This Book On Your TV

5. Plan Ahead
6. Put a Priority On People

1. PRAY FOR GOD'S HELP

Remember Philippians 2:5? "Let this mind be in you, which was also in Christ Jesus." Can you honestly say the things you watch are in the mind of Jesus? He is deeply concerned about the thoughts and images we fill *His* mind with. If you have been carelessly allowing every kind of pollutant to enter the gates of your eyes and ears, confess any poor TV habits to the Lord.

If your loved ones are also trapped by TV's snare, pray for them, too. Whatever you do, don't nag. Pray, instead, and let God do the convicting.

Remember, too, that our loving God has the days of our lives numbered. Ask yourself: Do I want to use that limited, God-given time in front of the tube?

2. PLUNGE IN

Plunge right in and acquaint yourself with the following ideas and your own "TV Alternatives." Open up your mind and pour in heavy doses of activity ideas which can be recalled when your TV whines for attention.

3. PERSONALIZE THIS BOOK

Make this list your own by highlighting the activities that are most appealing to you or someone in your family. There is no doubt that once you have broken the old pattern of automatic TV viewing, you will begin thinking of hundreds of new ways to spend your time. Jot down your ideas here in this book. Include ways to suit the activities to your particular family's needs. Note in the margin the title of a book that you've been meaning to read or a particular park that you want to hike through.

In some sections, blanks have been provided for you to fill in information such as phone numbers and daily hours of skating rinks and swimming pools. The "Indoor Games" and "Outdoor Games" sections have a place reserved for your personal game inventory. The more you personalize this book, the sharper it will become as a weapon against old habits of too much TV.

4. PLACE THIS BOOK ON YOUR TV
After you have read and highlighted the alternatives list, you may be wondering how you will ever remember the ideas when you need them. I suggest that you help your memory as well as your self-discipline by leaving this book in plain view on top of your television set. In those moments of temptation, before you flip on the TV, flip through the list of alternatives and find an activity which interests you.

5. PLAN AHEAD
Most of the ideas listed can be done on the spur of the moment. However, some of the activities require a little forethought and preparation. Don't let this scare you off. This preparation in itself can be an alternative to television. When I sit down to make my "To Do List," I've learned also to jot down a few fun activities that I want to prepare for myself or my family to do during the week. This may mean gathering supplies for a craft, writing up 3 x 5 cards for a verse memory game, or phoning some friends and inviting them to join our family on a particular evening. Sometimes a few minutes of forethought may be just what you need to make your non-television activities more attractive than Hollywood's allure.

6. PUT A PRIORITY ON PEOPLE
It is amazing how the simple addition of a few guests

turns an ordinary activity into a special event.

I remember an evening when my husband was out of town on a business trip. My daughters and I had enjoyed a lot of togetherness that day, but by evening I felt that I had honestly had all the preschool conversation that one adult could handle. I was determined not to waste our time on television, but the evening walk I had decided upon did not hold a great deal of appeal to me either; that is, until I thought about asking my neighbor to join us. She happily accepted my impromptu invitation, tucked her little ones into a stroller and we were off. As the wheels began to roll so did our conversation. Since that evening we have enjoyed many neighborly strolls together. I lost my sense of frustration while gaining a trusted friend.

At our house we have one more rule regarding people and TV: *People take priority over television.* In most circumstances, we insist that the TV goes off when company comes in. Giving our full attention to others is one of the best ways we can show them that they are valuable. Honor others by showing them that they are more important than even your favorite TV show.

CHAPTER ELEVEN

TV ALTERNATIVES

PART ONE: FAMILY NIGHT

Show your family that you enjoy spending time with them by declaring a weekly or even a monthly "Family Night." I define "Family Night" as an evening reserved exclusively for doing fun things with the immediate family. It is a great way to demonstrate that you value their company as well as a perfect opportunity to model what one can do other than watch the television.

Ideas listed in Part One are particularly well suited to Family Night activities, but consider scanning the alternatives listed under other headings for greater variety.

FAMILY NIGHT - at a glance

1. Plan a Cultural Dinner
2. Mail Bible Portions to Foreign Countries
3. Assemble a Missionary Care Package
4. Read Stories with a Message
5. Modernize and Dramatize a Familiar Bible Story
6. Produce Your Own Puppet Show
7. Record an Audio or Video Tape for a Friend or Relative

8. Talk About Memories and Record
9. Take Fun Pictures of Your Family
10. Look at Old Photos
11. Sing Together
12. Watch a Wholesome Video
13. Scripture Memory Verse Games
14. Picnic: Breakfast, Lunch, or Supper in the Park
15. Camp out in the Backyard

* * *

1. PLAN A CULTURAL DINNER

Plan an evening where your menu, dress, and conversation all focus on a specific foreign country. Check out the library for music recordings, videos, recipes, and interesting trivia about the culture you have chosen. Play one of the country's popular children's games. Pray for the leaders and Christians of the land. An excellent resource book to guide you is P.J. Johnstone's *Operation World* (available from any Christian bookstore). It lists statistics and prayer requests for every nation in the world.

* * * * *

2. MAIL BIBLE PORTIONS TO FOREIGN COUNTRIES

Put your love for God's Word into action by mailing Bibles to some of the four billion people of the world who have never owned a Bible. Organizations such as Bibles For The World will provide you with the Bibles, packaging, and labels. Even young children can participate in this simple yet rewarding project. All you do is wrap, mail, and pay for the postage. For a free 12-pak introductory Bible kit, call Bibles For The World at: (800) 323-2609. In Illinois, call (312) 668-7733.

* * * * *

3. ASSEMBLE A MISSIONARY CARE PACKAGE

Boost the morale of your missionary friends on the front lines of spiritual service by mailing them some items that they personally need or would just enjoy. Some ideas might be:
- Greeting cards and envelopes
- Special occasion paper napkins and tablecloths
- Birthday candles
- Chocolate chips
- Christian music tapes
- Christian and educational magazines for adults and children. (Send your own recent copies or buy some at your Christian bookstore.)

An extra note:

In some countries, missionaries have to pay duties of 100% or more of an item's value just to accept the package. Ask your missionary or missions board what you can do to avoid these expenses to your overseas friend. For instance, it has been suggested that one should send used rather than new books and mark the package accordingly.

* * * * *

4. READ STORIES WITH A MESSAGE

For entertaining and inspirational reading, borrow books and magazines from your friends or church library, or make

a family outing to your local Christian bookstore. Select stories from the following categories:
• Bible, Children's Bible, or Bible story books
• Novels that emphasize biblical principles
• Biographies of great Christians
• Missionary stories - From books, magazines and mission newsletters such as *In Other Words*, published by: Wycliffe Bible Translators, P.O. Box 2727, Huntington Beach, CA 92647

To "jazz up" a story. . .
Either have one reader read expressively to the whole family or take turns reading to each other. For a dramatic touch, have the reader dress up as the main story character. Or, the whole family could act out the story as it is narrated.

Older listeners can draw illustrations for the book being read while younger ones color them. Big brother and sister can also have fun illustrating the story for the little ones in the family with puppets or flannelgraph figures. (I either buy our figures at the educational or Christian bookstore or make them from our girls' coloring books and Sunday School papers. Our board was easily made by attaching flannel to a rectangular piece of cardboard.)

* * * * *

5. MODERNIZE AND DRAMATIZE A FAMILIAR BIBLE STORY
Look at an old, familiar story in a fresh, new light by placing yourself in the modern equivalent of Bible-time challenges. For instance, consider what modern obstacles Daniel might have faced if he had lived in today's society. Make a lasting

memory of your skit by videotaping it. Here are some suggested stories for you to dramatize.

Bible characters:
- Noah (Genesis 6—9)
- Mary and Joseph (Luke 1—2)
- Jonah (Jonah)
- Daniel (Daniel 6)
- Zacchaeus (Luke 19)

Parables:
- The Lost Sheep (Matthew 18:10-14)
- The Unmerciful Servant (Matthew 18:21-35)
- The Workers in the Vineyard (Matthew 20:1-16)

* * * * *

6. PRODUCE YOUR OWN PUPPET SHOW

A simple impromptu show can be wonderful, creative entertainment. Begin by making your own puppets out of paper bags, paper plates, popsicle sticks, socks, cardboard tubes, dolls, or stuffed animals. Then convert a cardboard box or just a cloth-covered table into a puppet stage. Now you can perform your own version of a nursery rhyme, children's story, Bible story, or your family's one-of-a-kind original script. If your family is really good, volunteer to perform in Children's Church. Leaders are always looking for a fresh way to relate a lesson.

* * * * *

7. RECORD AN AUDIO OR VIDEO TAPE FOR A FRIEND OR RELATIVE

Send that missed loved one a special message that they will never forget. Let your whole family individually share their news, sing a song or even show-off a little, and you will have a unique "letter" that will leave the listener with the feeling that they have practically been with you. Below are some ideas for making your tapes more interesting.

- Don't feel compelled to fill the entire tape. Simply stop when you are done.
- Enhance your audio tape with humorous sound effects.
- Record your video in more than one location.
- Make family members magically appear by photographing first an empty sofa, then pause the camera; have one member sit down on the couch and start the camera again. Repeat these steps until each member seems to have miraculously appeared on the sofa.
- Have children (or adults!) perform short stunts like turning cartwheels, swinging from a tree, playing a segment of a piano piece, making a great football pass, or belting out a cheer routine.

* * * * *

8. TALK ABOUT MEMORIES AND RECORD

If you never seem to get around to writing down your family history or those especially comical experiences, making a tape may be the easy way for you to record those precious moments. Invite an older relative to join you and share stories from earlier years. Remember that youngsters can also contribute and will immensely enjoy reviewing the tape in years to come. Begin your family's memory tape by selecting a topic from below.

- What are some funny things the kids did when they were little?
- Who made the greatest impact on your life and why? Describe them.
- How and where did you become engaged?
- What were the events surrounding your salvation?
- What were some of the highlights of your school years?
- What did you do during the war (for older relatives)?
- How would you describe one of your favorite times with Grandpa (or another relative)?
- What was your favorite Christmas or birthday? Why does it stand out in your memory?

Now just turn on the recorder and give each person an opportunity to respond to the question. When you have exhausted a subject, label the tape and store it for your next recording session.

* * * * *

9. TAKE FUN PICTURES OF YOUR FAMILY

Are the years slipping by without a visual record for your family's memory? Take the time now to photograph the people who are most important to you. Be silly; be creative. Photograph family members holding their most cherished possessions. Have your young children model in adult clothes. Think of a theme such as your favorite vacation spot, either one you've been to or one you'd like to visit. Have everyone dress for that vacation. Take pictures of individuals such as dad repairing a bike in the workroom. (*I wish that I had a photograph of my childhood home, the kitchen, my bedroom, our spacious attic. There is so much that we forget and photographs renew our slipping memories.*)

* * * * *

10. LOOK AT OLD PHOTOS
Every once in a while a family should pull down the photo albums, home videos, slides, or movies and relive some of their family history.

* * * * *

11. SING TOGETHER
Sing for fun and don't worry if you don't sound perfect. A hymnbook will aid your impromptu concerts. Let your children teach you some new songs from the youth group or children's choir. You may even want to select a number to practice for a performance at a family gathering or church. If no one plays an instrument at your house, either sing *a capella* as we do, or sing along to an instrumental tape.
 • Sing before a nightly story or devotional time.
 • Sing while taking a hike in the woods.
 • Sing while riding together in the car. (*We sing on the*

way to church. It calms us after the rush of getting ready and reminds us that the day belongs to the Lord.)

- Sing before you give thanks at the evening meal. (*In our home this is a tradition that focuses our attention and sets a positive tone for the evening.*)

* * * * *

12. WATCH A WHOLESOME VIDEO

Transform your video night into an "evening of entertainment" for your family and a few guests. Let the kids set up and run a concession stand. Turn off all the lights to enhance the atmosphere, and have a child use a flashlight to usher "patrons" into the darkened room. Here are some good video choices:

- **Christian videos:** The video review section of this book has over 100 suggestions.
- **Family shows:** Musicals, classics, animated stories, and comedies are good choices. Many of these can be prerecorded from the TV for use on your special evening.
- **Home videos, movies, and slides:** These can be as much fun as any commercial production.

* * * * *

13. SCRIPTURE MEMORY VERSE GAMES

Make up your own or try these. They're fun as well as educational.

Scramble

1. Write each word of the Bible verse on a separate 3 x 5 card.
2. Mix up the cards.

3. Try to put the cards back in the correct order.
4. For stiffer competition, time how long each person takes to put the cards in order.

Tick-Tack-Toe
1. Write the Bible verse references on cards.
2. Place cards face down on the table.
3. Teams take turns drawing a card and reciting a verse.
4. Members who recite the verse correctly may place an X or an O in the tick-tack-toe square.
5. Make the game easier by putting the first word in the verse along with the reference.

Match up
1. Write verses on 3 x 5 cards and spread the cards out face up on a table.
2. Write the references to the verses on other 3 x 5 cards and place face down in a stack on the table.
3. Players take turns drawing reference cards and matching with verse cards.
4. Player keeps verse cards for each correct match.
5. Winner is the player with the most cards at the end of game.

* * * * *

14. PICNIC: BREAKFAST, LUNCH, OR SUPPER IN THE PARK

Picnics are a tried and true winner with people everywhere. Consider these variations on an old theme.
• Serve the meal on china complete with goblets of sparkling grape juice.
• Pick a theme like "Hillbilly Picnic." Wear clothes and

111

play games that fit the part.
- Plan a breakfast picnic. Bicycle to the park on a Saturday morning and enjoy breakfast in your favorite uncrowded spot. Include a devotional reading and prayer time. (*Some of my favorite childhood memories were made in this very fashion.*)

* * * * *

15. CAMP OUT IN THE BACKYARD

Begin your backyard adventure with a hobo dinner (sliced potatoes, carrots, onions and ground beef cooked in an envelope of foil) on a grill or bonfire. Set up a tent; pull out the sleeping bags, a lantern, and a big bag of popcorn. Then settle down for a family time of singing, storytelling, and stargazing.

* * * * *

PART TWO: GOING AND DOING

This is a list of things to do that will get you out of the house. The new surroundings may be just what you need for a fresh attitude and a relaxed feeling. Under several of the headings, space has been provided for you to write the names, phone numbers, admission fees, and business hours of local attractions. These quick reference helps will make scheduling your outings fast and convenient.

GOING AND DOING - at a glance

16. Take a Bike Ride
17. Feed the Ducks
18. Explore Your Library
19. Take a Hike
20. Go Roller Skating
21. Go Ice Skating
22. Attend a Concert
23. Eat Salad and/or Dessert at an Exclusive Restaurant
24. Visit a Museum or Planetarium
25. Go Bowling
26. Visit the Zoo
27. Go For a Sunset Drive
28. Go Swimming
29. Go Horseback Riding
30. Go Out for a Frozen Dessert
31. Go on a Weekend Campout
32. Leave an Anonymous Gift at a Friend's Door
33. Visit Your Chamber of Commerce
34. Sit in on a City Council Meeting, Visit a Courtroom In Session or Your State Legislature
35. List Attractions Available in Your Area

16. TAKE A BIKE RIDE

Each year is filled with gorgeous days that we simply let slip by. Don't let another one pass. Try a devotional cruise for just you and the Lord. Recite familiar Scriptures, or spend your time talking to God. Plan to stop at a friend's house for a short "hello." Bring an extra smile with cookies or flowers from your garden as a gift of friendship.

Kids only:

- **Races:** Invite your friends to join you for speed races in a deserted parking lot. Try slow races, too.
- **Follow the Leader:** Play "Follow the Leader" on bikes. Pretend that you are biking on a safari or in a foreign country.
- **Bike Parade:** Have your friends join you in decorating your bikes outlandishly. Weave streamers through the spokes of your bike. Create a motorized sound by clothespinning a playing card to the spokes that will flip as the wheels turn. Maybe you will want to go all out and disguise your bike with big pieces of cardboard to look like a car, a carriage, or even a dinosaur.

* * * * *

17. FEED THE DUCKS

An outing as simple as a trip to the park can be the backdrop for a precious memory. Call an elderly neighbor and invite him/her to join you in spreading a feast for hungry birds. Pack up your stale bread, popcorn, and cereal but don't forget the camera so you can capture the outing on film.

18. EXPLORE YOUR LIBRARY

Make this trip to the library an adventure. Scan the magazines your library offers. There are magazines about parenting, cooking, cars, photography, sports, writing, teaching, business, antiques, dogs, cats, fish, films, four-wheeling, skydiving and almost every other special interest imaginable. Let your second discovery be the audio-visual department and its array of records, audio tapes, video-cassettes and films.

19. TAKE A HIKE

Get in touch with nature in your favorite state park or national forest. Before you go, check out books from the library on native state plants and animals. Pack a lunch, water, a can of insect repellent and some bandaids for your trek. Select your clothes with the weather in mind. Take your camera. Park officials can point out landmarks and provide you with maps. Consider joining one of the park's planned hikes. As you hike, look for things that reflect the Creator and share these thoughts with one another.

20. GO ROLLER SKATING

Whether you choose street or rink roller skating, it is a great way to exercise and have fun, too! Outdoor basketball courts are a great place to skate, but my personal favorite is a big deserted parking lot. I used to get my exercise by rising early to skate around the mall parking lot before the stores opened. If you are skating with a friend, play tag; see if you can dribble a ball while skating. Bring along a tape player and skate to music.

Call your closest rink to find out what nights they play Christian music. Then get on the phone and invite friends to join you. You might plan a clothing theme like "crazy ties" or "silly socks." Be friendly and decide that everyone has to meet at least one person who is not with your group before the evening is over. You can have fun turning a few heads by agreeing that when one person calls out "crazy ties!" everyone converges in the center of the rink, shakes hands, and then returns to nonchalantly skating around the rink as before.

Quick reference information:

Name of rink _____

Address _____ Phone #: _____

Cost _____ Hours _____

* * * * *

21. GO ICE SKATING

During the cold months you can take advantage of rinks and frozen ponds. Always check the depth of the pond ice and bring along a rope or pole for emergencies. Some ideas: Play hockey with brooms and a homemade puck made from a butter tub filled

with gravel and secured shut with strapping tape. Or, perform your own mock "ice capades." Don't forget about ice skating in the summer, either. Spending an afternoon at an indoor rink is a great way to beat the heat!

Quick reference information:

Name of rink _____

Address _____ Phone #: _____

Cost _____ Hours _____

* * * * *

22. ATTEND A CONCERT

Enjoying a great concert doesn't have to cost a lot of money. Check your newspaper for city or church activity listings. Frequently these concerts are free, but very good. Sometime, though, you might want to go all out, purchase tickets, and dress up for an evening at the symphony.

* * * * *

23. EAT SALAD AND DESSERT AT AN EXCLUSIVE RESTAURANT

Dress up for a "meal" out at an exclusive restaurant—without killing your budget. Eat a larger than usual lunch at home, and then simply order salad and dessert. Chances are you will probably get a nice basket of bakery rolls, too. It can be a quiet date for two, or an opportunity to invite another couple to join you. Parents can make this a date with a child.

24. VISIT A MUSEUM OR PLANETARIUM

This is a fun as well as educational activity that will give you a new appreciation for art, history, science, or nature. If the museum has a current, special exhibit, read up on it a bit before you make your visit. Keep in mind that secular museums and planetariums are steeped in evolution which they present as a proven fact rather than a theory.

Quick reference information:

Name of museum _____

Address _____ Phone #: _____

Cost _____ Hours _____

Name of planetarium_____

Address _____ Phone #: _____

Cost _____ Hours _____

* * * * *

25. GO BOWLING

Relax a little and make the evening fun. See how many heads you can turn with some ridiculously elaborate wind-ups performed with stoic expression. Once the game gets rolling, you might enjoy a little friendly competition by declaring a booby prize for the losers such as opening all the doors for the winning team that evening.

Name of bowling alley _____

Address _____

Phone #: _____ Hours _____

Cost: per game _____ shoe rental_____

* * * * *

26. VISIT THE ZOO

Revive this old favorite by giving each person in your group a list of things to look for in the zoo such as: a nocturnal creature, an animal that lives in your state, a water bird, an animal from Australia, an animal from India, etc. The first to find all the answers gets a snowcone (or some other treat). On another day at the zoo bring along your camera and take comical pictures of your friends or family imitating the animals in their cages. When the photos are developed, you can make up an album called, "Bizarre and Unusual Creatures."

Quick reference information:

Zoo name_____

Address _____ Phone #: _____

Cost _____ Hours _____

* * * * *

119

27. GO FOR A SUNSET DRIVE

Take a relaxing drive to a scenic spot and watch the sun set.
Then, stop for a soda on your way home.

* * * * *

28. GO SWIMMING

Splashing in the water is a great way to cool off in the
summer heat. Whether you swim in a wading pool or the
ocean, water games will make the day even more fun.

Water games
- Retrieve a coin from the floor of the pool or lake.
- Play tag.
- Play catch or keep-away.
- Stand on your hands on pool bottom.
- Play water volleyball or basketball.
- See who can make the biggest splash.
- Have your group stand in a circle and toss a beach ball
  back and forth, always trying to keep it in the air.
- Have swimming races or running-in-the-water races.

Keep your swimming safe by following these important
rules:
1. *Never* leave a child unsupervised by the water.
2. Don't dive where the water is shallow or there are any
   obstructions.
3. Insist on a buddy system for a large group.

Quick reference information:

Public pool name_____

Address _____ Phone #:_____

Cost _____ Hours _____

* * * * *

29. GO HORSEBACK RIDING

A special trip to the stables is definitely on the memory-maker list. Be sure to bring the camera and record this fun outing for family memory night.

Quick reference information:

Stable name _____

Address _____ Phone #: _____

Cost _____ Hours _____

* * * * *

30. GO OUT FOR A FROZEN DESSERT

This mini-outing is a favorite at our house. More than once we have been seen pulling our little red wagon loaded with kids up to the corner yogurt shop. The evening stroll and after-dinner treat are the perfect opportunity to hold hands and talk, something I never get enough of.

* * * * *

31. GO ON A WEEKEND CAMPOUT

Get completely away from the everyday routine with a weekend campout.

Planning will prevent most frustrations. Joining a family of veteran campers might be helpful for your first trip out. Decide ahead of time your route, destination, supplies, and schedule. Then determine to be flexible. As you pack your camping gear, include these items: a ball, frisbee, board games, pencil and paper, bags for collecting woodsy treasures, a magnifying glass, a Bible, songbooks, camera.

(In their excellent book The Blessing, *Gary Smalley and John Trent printed the response of people in numerous seminars and counseling sessions to the question: "What is one specific way you knew that you had received your parents' blessing?" It was interesting to read that one of the answers which they received most often was: "We went camping as a family.")*

* * * * *

32. LEAVE AN ANONYMOUS GIFT AT A FRIEND'S DOOR

Take some time to brighten a friend's day with an anonymous gift. You could give a favorite dessert, a cartoon book, or a restaurant gift certificate. Deliver the gift to their front doorstep, ring the doorbell, and watch the surprise from behind the bushes.

* * * * *

33. VISIT YOUR LOCAL CHAMBER OF COMMERCE

They may have more ideas of local places of interest or ways to become involved in your community.

Street address _____

Phone #: _____ Hours _____

* * * * *

34. SIT IN ON A CITY COUNCIL MEETING, VISIT A COURTROOM IN SESSION OR YOUR STATE LEGISLATURE

This activity is interesting and educational for all members of the family.

City Council meets _____

Street address _____

Time _____

Trial or session time _____

State Legislature meets _____

Our local representative is _____

* * * * *

35. LIST THE ATTRACTIONS AVAILABLE IN YOUR AREA

Use this space to list attractions in your area that you would like to visit someday. Then, when you don't know what to do, you can quickly scan your list for appropriate activities.

_____

_____

_____

_____

_____

_____

* * * * *

PART THREE: INDOOR GAMES

These are some reminders of what you can play inside the house, just for the fun of it. The ideas listed cover a wide range of ages. Most of them can be enjoyed by anyone who is young at heart. We encourage parents to take some time to play with their children, but, kids can use this list to make their own fun, too.

INDOOR GAMES - at a glance

36. Hide and Seek
37. Board Games and Card Games
38. Jigsaw Puzzles
39. Crossword Puzzles
40. Forgotten Toys
41. Building Block City
42. Domino Trails
43. Ping Pong
44. Balloon Games
45. Water Glass Concert
46. Paper Airplanes
47. Rhyming Words
48. Block Bowling
49. Charades
50. Tent Hideouts
51. Aroma Bag
52. Twenty Questions
53. Name That Sound
54. Pretend

* * *

36. HIDE AND SEEK
In the original game, one seeker hunts several hidden children. Try reversing this by choosing one hider and several seekers. Make it even more fun by declaring that a seeker who finds the hider joins him in his hiding spot until all are discovered by the last seeker.

Another variation of the game is to hide an object. Give the seeker clues by saying, "You are getting hot," or "You are getting cold." Finding their hidden toys is a delight for small children, and they are anxious to play again and again.

* * * * *

37. BOARD GAMES AND CARD GAMES

Take an inventory of your family's games and write them here for quick reference in the future. Then, you may want to make a trip to a toy store, department store, or educational store to pick out a few additional games that your family would enjoy. Don't forget to watch garage sales for bargains on games, too. Or, you might swap games with a friend for a month. Here are a few suggestions for your consideration.

Table games we have   *Table games to buy*

_____   _____

_____   _____

_____   _____

_____   _____

_____   _____

_____   _____

Other games
- Password
- Dictionary
- Concentration
- Pick-up sticks
- Jacks
- Darts
- Nerf basketball
- Ring Toss
- Matching games for colors, letters, or pictures
- Dominoes (look for variations like animal and alphabet dominoes)
- Bingo (look for educational versions that emphasize math or language skills)

* * * * *

38. JIGSAW PUZZLES
Jigsaw puzzles can be lots of fun alone or in groups. The giant puzzles may take days to complete, so assemble them on a board, card table, or some other surface that can be easily moved out of the way when not in use.

* * * * *

39. CROSSWORD PUZZLES
You can develop your vocabulary and have fun at the same time by completing crossword puzzles. Find puzzles in books, the daily newspaper, and children's Sunday School papers. Clip puzzles out of papers whenever you see them and store in a shoebox for a TV alternative some night. However, you might find that you can have just as much fun challenging others with your own homemade puzzles.

* * * * *

40. FORGOTTEN TOYS

Select a few toys that seem to have lost their appeal and store them out of sight for a few months. Later, pull them out to be played with. You might be surprised at your children's interest in these "old" toys. Occasionally, it will be necessary to sort out the toys your children have outgrown. Suggest that your child select several of these toys to be given to friends or organizations which will appreciate them. Take your child with you so that they can give the toys to others themselves.

* * * * *

41. BUILDING A BLOCK CITY

Every age enjoys building. Just vary your techniques and your building materials and you have an activity suited to any ability level. Some of the materials you can use are: wooden blocks, Legos, Tinker Toys, Lincoln Logs, cardboard tubes, boxes, and even sheets of paper which can be folded, taped, and colored. There is an unlimited number of structures that can be built. Here are some kick-off ideas to get your young engineers and architects started.

- airport
- church
- dog house
- doll house & furniture
- farm
- ferris wheel
- lawn mower
- skyscraper
- space station
- submarine
- train
- _____

* * * * *

42. DOMINO TRAILS
Line up dominoes on end and in trails or patterns across the floor. When you're finished, hit the leading domino and watch them fall. You can make forks in the path which set off two or three other paths, make bridges of books or blocks. Go under tables and around chair legs. Time how long it takes for your trails to collapse. Compete with yourself for longer collapsing times. Before you set off your most spectacular trails, pull out the camera or video camera.

* * * * *

43. PING PONG
Challenge your friends to a game of ping pong or begin a whole tournament. Maybe you will want to create your own games or enjoy an old variation like Around the World (also known as Round Robin).

* * * * *

44. BALLOON GAMES
Here are some silly ideas for how to play with a balloon.

Balloon bombs
This is an active game for one person or a whole crowd. Blow up four or five balloons. The object of the game is to keep the balloons in the air at all times. Pretend that if it touches the ground or the furniture it explodes.

Clinging balloons
Rub balloons on your hair to create static electricity. Find out what your balloon will cling to. Look at yourself in the mirror. Is your hair sticking up?

Balloon squash
Race other players to see who can sit on and break the most balloons.

Balloon toss
How far can you toss a balloon? Compete for distance.

Balloon people
Draw faces on balloons. Cut shoes, ears, hats, etc. out of paper and tape to balloon faces. Create balloon portraits of your family.

*Caution
When playing with balloons, remember that broken balloon pieces can be swallowed and choked on by little ones. Be alert and pick up all broken pieces.

* * * * *

45. WATER GLASS CONCERT
Create your own "percussion instrument" by filling drinking glasses with varying amounts of water. A water glass is "played" by gently tapping the side with a spoon. "Wind instruments" are made by filling narrow-necked bottles with varying amounts of water. Play these by blowing across

the top of the bottle opening. Experiment with different water amounts until you can perform your own very *unique* concert.

* * * * *

46. PAPER AIRPLANES
Let your creativity fly with paper airplanes. Experiment with various designs and sizes as you compete for flight distance. There are some great library books on how to make paper airplanes and other things that fly. Look under the Dewey Decimal call numbers 745.592. I recommend these two books!

Air Crafts: Playthings to Make and Fly, by Leslie Linsley and Jon Aron
Easy-to Make Spaceships That Really Fly, by Mary Blocksma and Dewey Blocksma

* * * * *

47. RHYMING WORDS
The starting player chooses a good, short word for rhyming such as: hat, can, or night. Then, everyone gets 60 seconds to write down words that rhyme with the starting word. The player who has the most correct rhyming words wins the game.

* * * * *

48. BLOCK BOWLING

Yes, you *can* go bowling right in your family room! It is just a matter of playing with old toys in a new way. Set up tall building blocks on end in a triangular pattern. Station players across the room and roll a ball at the blocks. Give points for the number of "pins" knocked down on each roll.

* * * * *

49. CHARADES

This time-tested game is always good for a few laughs. Silently act out your clues so other players can guess the book, movie, or song title that you are dramatizing. Give younger children different categories: ordinary activities (like sweeping or brushing teeth), animals (a monkey, an alligator), Bible stories (like Jesus calming the sea) or familiar children's titles to dramatize.

* * * * *

50. TENT HIDEOUTS

Create tent hideouts by draping sheets and blankets over tables and chairs. Secure the sheets in place with pins and heavy books, or by closing the sheet corners in drawers.

* * * * *

51. AROMA BAG

Choose an item with a distinct smell such as an apple, onion, or a perfumed hanky and put it in a paper bag. Blindfolded children then take a whiff through the opening of the bag and make a guess at what is inside.

* * * * *

52. TWENTY QUESTIONS
This game for two or more can be played anytime, anywhere. To begin, the first player thinks of a person, place, or thing. Then the other players ask him questions to determine what he is thinking about. Only questions which can be answered with a "yes" or a "no" may be asked. Continue gathering clues with up to twenty questions until the secret identity is unveiled.

* * * * *

53. NAME THAT SOUND
These are simple guessing games for identifying sounds.

Game one
Tape record various noises throughout the house such as running water, the clatter of dishes being put away in the cupboard or the padding of footsteps. Then, playback the tape for others to hear and name the sounds. This can also be expanded by taping other sounds throughout your locale such as a train, cows mooing, traffic in a tunnel, and so on.

Game two
Play only the beginning notes of a familiar song. See if others can identify the tune.

* * * * *

54. PRETEND

When you pretend, you can go anywhere, be anything, and do everything! Parents can teach as well as learn a lot about their children by watching or, better yet, participating in their children's pretend play. Below are examples of the endless possibilities for pretend play. Add your own. Make up stories; act out the adventures. Have fun!

Adventures
- jungle safari
- space exploration
- underwater treasure hunt
- shipwrecked on a desert island
- _____

- _____

Feelings
- winning or losing a contest
- meeting a new friend, losing an old one
- getting a new baby
- moving to a new place
- _____

- _____

Occupations
- ballerina
- fire fighter
- missionary
- _____

- doctor
- lawyer
- Indian chief
- _____

134

PART FOUR: OUTDOOR GAMES

One of the best things about outdoor games is that they combine fun with the exercise which we know we should be getting, let off steam, and get some fresh air all at the same time. Many of these games can be enjoyed by all ages right in your backyard.

OUTDOOR GAMES - at a glance

55. Frisbee
56. Play Ball
57. Water-Balloon Toss
58. Kick-the-Can
59. Tug-of-War
60. Yard Games That Use Special Equipment
61. Fly a Kite
62. Sledding, Sculptures, and Snow Forts
63. Hopscotch
64. Good Shepherd
65. London Bridge
66. Blow Bubbles

* * *

55. FRISBEE
There are lots of ways to play with a frisbee. You can play "fetch" with the dog. You can have a good time playing a simple game of catch or keep-away with your friends, or you can try a few of the games described here.

Frisbee Football: Substitute the frisbee for the football and a toss for the football's kick-off and then play by basically the same rules as regular football.

Frisbee Golf: Buckets or garbage cans replace the holes in golf as you try to toss a hole-in-one. Keep track of the number of tosses it takes to complete the course. The lowest score wins.

Target practice: Aim at trees, fences, or even tin cans to improve your accuracy.

Check out a book from the library on the subject of frisbee games. I recommend, *Frisbee: More Than A Game Of Catch* by Judy Horowitz, Women's World Frisbee champion.

* * * * *

56. PLAY BALL
Whether you play the full game or just practice your dribbling, passing, and throwing, this is a fun way to get your exercise. Don't forget these old favorites.

- Basketball
- Catch
- Football
- Four-Square
- Handball
- Kickball
- Softball
- Volleyball

* * * * *

57. WATER-BALLOON TOSS
To play this game you need at least four people (two teams

of two each). Each team begins by tossing a water balloon to one partner. After each successful catch, each of you should take a step backward and throw again. If the balloon bursts, your team is eliminated. The last two players left in the game are the winners.

* * * * *

58. KICK-THE-CAN
Kick-the-Can is played by two or more kids using only their feet to kick or steal an empty tin can in an active game of keep-away.

* * * * *

59. TUG-OF-WAR
A familiar game in which the center of a sturdy rope is marked and teams take their places at opposite ends. On "Go!" each team pulls and tugs on the rope in an effort to make their opponent cross the center line. You can add some extra excitement to the game by placing a wading pool or even a mud puddle between the two teams. In this case, the losers will be obvious!

* * * * *

60. YARD GAMES THAT USE SPECIAL EQUIPMENT
Below are a few good yard games which, if you don't already own, you might consider buying. Use the blanks to make a list of your own family's yard games for future reminders.

- Badminton
- Croquet
- _____
- _____

- Horseshoes
- Tetherball
- _____
- _____

* * * * *

61. FLY A KITE

An ever-popular pastime. Find some open spaces and have a great time sailing your kite.

* * * * *

62. SLEDDING, SCULPTURES, AND SNOW FORTS

If you live in the right part of the country at the right time of the year, there is no end to the possibilities for snow play. You can sled with the side of a refrigerator box, design a park bench in the snow, sculpt a snow dinosaur in your front yard, or even build a snow house. Then, prepare for snowball attacks by building a walled fortress.

* * * * *

63. HOPSCOTCH

An old-fashioned game that today's high-tech kids may not know very well. All you need is a piece of chalk to draw your hopscotch outline on the driveway and a stone for each player. Begin the game by tossing your rock onto the square numbered 1. Then

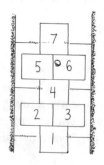

hop on one foot in each square except the one which has a rock in it. When you get to the end, turn around and start hopping back. When you reach the square with your rock, pick it up and hop on out. The game continues as you roll your rock to box 2, 3, 4. You lose a turn if you:
- step on a line
- step on a square with a rock in it
- put two feet in one square
- your rock rolls into the wrong square.

The winner is the first person to move their stone all the way to number eight.

* * * * *

64. GOOD SHEPHERD

Toddlers are more successful at this variation of "Simon Says" called, "Good Shepherd, Bad Shepherd." We tell our toddlers, "The Good Shepherd says scratch the back of the person next to you. The Bad Shepherd says make a mean face." The goal is to be a good listener and only do what the Good Shepherd says and never what the Bad Shepherd commands. It's a good reminder for us to listen for the voice of our one, true Good Shepherd.

* * * * *

65. LONDON BRIDGE

As you probably remember, this game is played when two children clasp hands and hold up their arms to form a "bridge" under which the other children walk single file. Instead of the traditional song of London Bridge, try new words such as:

"Jesus Christ loves you and me,
you and me,
you and me.
Jesus Christ loves you and me.
And we love you, too."
At this point the arms of the bridge collapse capturing whoever was walking under it. The song continues,
"We'll give you a hug and kiss,
hug and kiss,
hug and kiss.
We'll give you a hug and kiss,
because we love you so."
After receiving a hug and a kiss, the child is released and the game continues as before.

* * * * *

66. BLOW BUBBLES

Enjoy the wonderful colors and sizes of these floating, exploding delights. To make your own bubble solution, mix four parts water and one part liquid dishwashing soap. You can also add a drop of food coloring for new colors. Many items around your house make great bubble wands such as drinking straws, pint-sized plastic fruit crates, or tin cans with the top and bottom removed.

* * * * *

PART FIVE: IN THE KITCHEN

It has been said that the kitchen is the heart of the home. Welcome your friends and family members into your kitchen—and your heart—with creative, taste-tempting treats. Whether you are doing dishes or preparing some special treat together, time in the kitchen might be just the right ingredient for stirring up the best conversations and warmest memories. The TV Alternatives listed here will start you cooking with fun activities that will get you out of the TV room and into the kitchen.

IN THE KITCHEN - at a glance

67. Make Caramel Corn
68. Make Pretzels
69. Make Caramel Apples
70. Make Gelatin Shapes
71. Bake and Decorate Cookies
72. Kids: Plan and Prepare a meal

* * *

67. MAKE CARAMEL CORN

Make popcorn and put it and peanuts in a roaster pan.
        16 cups (4 quarts) of popped corn
        2 cups salted peanuts

In a saucepan stir together:
        1/4 c. white corn syrup   1/2 c. margarine or butter
        1 c. brown sugar   1/2 t. salt
Bring these ingredients to a boil and cook five minutes. Remove from heat.

Stir in: 1/4 t. baking soda

Pour the mixture over the popcorn and peanuts. Stir it well. Bake uncovered at 250 degrees for one hour. Stir the baking mixture at 15 minute intervals. Cool and break apart the caramel corn on cookie sheets.

Munch your homemade caramel corn now, freeze some for later, or bag it up as gifts for your friends.

* * * * *

68. MAKE PRETZELS

In a big bowl, mix together:
    1 pkg. yeast
    1-1/2 c. warm water
    1 T. sugar
    1 t. salt
Stir in:
    4 c. flour

Knead the dough until it is smooth and then shape into traditional pretzel-shaped ropes, or be creative and make pretzel elephants, flowers and people. Brush your creations with beaten egg and sprinkle with salt. Bake at 450 degrees until browned, about 15 minutes.

* * * * *

69. MAKE CARAMEL APPLES

At our house, this is Dad's specialty. A couple of times a year, he and the children go shopping for the ingredients and then spend the evening preparing and enjoying their special treat together.

All you need is five apples, caramel pieces, popsicle sticks, and a little water. Most packages of caramels contain the sticks for your apples and have a recipe printed on the package as well. Insert sticks into the stem top of the washed apples and dip into the caramel sauce to coat. Place on greased wax paper to cool in the refrigerator.

* * * * *

70. MAKE GELATIN SHAPES

A healthful and easy treat is gelatin. Cut it into squares or holiday shapes you can pick up with your fingers to eat. You will need:

- 4 envelopes Knox Unflavored Gelatin
- 4 cups cold fruit juice (such as apple, cranberry, orange juice drink, lemon-lime)

In medium saucepan, sprinkle unflavored gelatin over 1 cup juice; let stand 1 minute. Stir over low heat until gelatin is completely dissolved, about 3 minutes. Stir in remaining 3 cups juice. Pour into 9-inch square baking pan; chill until firm, about 3 hours. To serve, cut into 2-inch squares or press cookie cutter shapes into pan. Remove carefully with thin, flexible spatula. Makes about 9 treats.

* * * * *

71. BAKE AND DECORATE COOKIES

Any cookbook has dozens of cookie recipes. Make a list of your favorites and make two special events out of this one. One would be a trip to the store to buy the ingredients; the other is the evening you bake - and *eat!*

* * * * *

72. KIDS: PLAN AND PREPARE A MEAL

Arrange a night once a month to be Kid's Night in the Kitchen. Kids may need some guidance initially, but eventually should be able to plan and prepare a meal for the whole family. Here are some guidelines that will help the event run more smoothly:

1. Include at least one food item from each of these categories: protein, green vegetable, starch, and fruit.
2. Give your grocery list to the family shopper at least two days prior to the meal.
3. Clean up the kitchen after the meal.

Make the evening more fun by using some of these creative ideas: Write and decorate a card that lists "Tonight's Menu" for your "guests." Play background music and decorate the table with candles and flowers. Wear a chef's hat and apron. Be creative as well as responsible, and Kid's Night is sure to become a favorite of the whole family.

* * * * *

PART SIX: READING AND WRITING

If you are struggling to overcome the TV temptation, prepare for success by placing your favorite reading material, paper and pen next to your easy chair. Promise yourself that you will not turn on the tube until you have read or written at least a page.

Reading and writing possibilities are limitless. However, this list is intended to give you a few specific ideas.

READING AND WRITING - at a glance

73. Read One of Your Dad's Favorite Books
74. Read to Your Family or Friends
75. Write a Love Letter
76. Cut and Paste a Funny Letter to a Friend
77. Write Letters to Your Elected Officials
78. Encourage Your Pastor with a Note
79. Write a Poem or a "Rap" About Your Children or Parents
80. Write a Comical Message for a Phone Answering Machine
81. Write a Short Story for Publication
82. Kids: Write and Illustrate Your Own Book
83. Write to a Prisoner

* * *

73. READ ONE OF YOUR DAD'S FAVORITE BOOKS

Sit down with a book that your dad once read and loved. You may learn something about him as well as honor him by showing an interest in a book he enjoys. And, who knows? It may become one of your favorites, too.

145

* * * * *

74. READ TO YOUR FAMILY OR FRIENDS

Share a good story with someone else. Even if it just means reading the paper to your wife while she is cooking dinner in the kitchen. Children can take turns reading out loud to brothers and sisters. Enjoy the story over several weeks by reading only a chapter a night. *My husband and I began our attack on the TV-habit by reading Christian novels to each other.* It's a great way to pass a cold winter's night in front of the fireplace, too.

* * * * *

75. WRITE A LOVE LETTER

Include specific qualities that you appreciate about your spouse or friend. Make your writing sincere. Use imagery from creation or every-day life to paint word-pictures that will give your message more meaning and beauty. The very fact that you took the time to write about your love or friendship for her (or him) will be a real encouragement. Mail your letter to the office, or hide it under the pillow.

* * * * *

76. CUT & PASTE A FUNNY LETTER TO A FRIEND

Do you have a pile of old newspapers and magazines sitting around somewhere? Pull them out along with scissors, glue, and a piece of paper. Snip and paste phrases and

146

words to create your own zany messages. If you can't find the right word you want, cut out a picture to represent it and include it in your sentence. Your creation is bound to be as much fun in the making as in the reading. Sign your name, or only hint at your identity. Then mail your letter to a friend who has a sense of humor.

* * * * *

77. WRITE LETTERS TO YOUR ELECTED OFFICIALS

Pen a note of encouragement to your federal or state representatives, your local school superintendent, or city councilperson. Praise them for positive voting records or even a street improvement if appropriate. Let your voice be heard on issues such as abortion, pornography, evolution taught in schools, or some other issue of importance. Officials will be much more able to meet the needs of the people if they know what those needs are.

* * * * *

78. ENCOURAGE YOUR PASTOR WITH A NOTE

As shepherd of the flock, your pastor probably receives many divisive, disturbing comments and "suggestions." It is difficult to be watched constantly and evaluated by others. Take a moment to minister encouragement to him or his family by writing a short letter of praise and assurance of your prayers for them.

* * * * *

79. WRITE A POEM OR A "RAP" ABOUT YOUR CHILDREN OR PARENTS

Make up a fun rhyme that describes each of your children, friends, or parents and expresses your feelings for them. Repeat it often and it will become a sweet and silly reminder of your love for them. Here is a "rap" that I wrote for one of our daughters:

> My little Sunshine Girl
>   is sweet Kristin Marie.
> She can make her dress twirl
>   and she loves her Mommy.
> We know she likes to tease
>   to laugh and run and play.
> I will give her a squeeze
>   and love her every day.

A poem for: _____

* * * * *

80. WRITE A COMICAL MESSAGE FOR A PHONE ANSWERING MACHINE

If you have an answering machine, have the family gather around the kitchen table and have each member write out a message for the recorder. After each has designed his or her personalized message, keep them. Change your recording each week as each family member records his/her message. Or, you may want to make a family message by entertaining callers with accents, sound effects, or special themes (English mansion, space ship, Old West, etc.). Be creative and have fun. Also, don't be surprised if your friends begin calling just to hear your messages!

* * * * *

81. WRITE A SHORT STORY FOR PUBLICATION

Try out your writing skills by preparing a short story, article, or filler for publication. You may have some experiences or knowledge that others can benefit from. Submit your story or article to your favorite magazines or Sunday School take-home papers. *The Writer's Market* by Writer's Digest Books, updated and published annually by F & W Publications, Cincinnati, Ohio can be found in most libraries. It is available from most libraries and provides information on where and how to get your work published.

I want to write a _____ for
_____ market.

* * * * *

149

82. JUST FOR KIDS: WRITE AND ILLUSTRATE YOUR OWN BOOK

Make up an adventure about someone like yourself and write it down. This is called the "rough draft" of your story. Read it over and make changes to improve it. Sketch some picture ideas for each page of your book. Now you are ready to neatly copy your story and pictures onto construction paper. Make a cover for your book from paper, cardboard, or cloth. Staple, sew, or glue the back edge together, and you have your very own originally written and illustrated masterpiece. Make one for Mom, Dad, or grandparents, and they will cherish it for a long time.

* * * * *

83. WRITE TO A PRISONER

Prison Fellowship Ministry, founded by Chuck Colson, offers volunteers an opportunity to write encouraging letters to inmates. Volunteers must be 21 years of age or have parental consent. To receive an application used to match you with an appropriate inmate, write to: Prison Fellowship Ministry, Pen Pal Program, P.O. Box 17500, Washington, DC 20041.

* * * * *

PART SEVEN: LISTENING AND LEARNING

The nice thing about just listening as opposed to the listening and *viewing* that TV requires is that our eyes and hands are free to engage in some other activity while we listen. If we spent more time listening, we would also probably learn more. Focus your mind on some of those things which you have always wanted to know, because when we stop learning, we stop growing. You are never too young or too old to listen and learn.

LISTENING AND LEARNING - at a glance

84. Listen to Music
85. Listen to Inspiring Messages
86. Listen to Others
87. Research Your Family Tree
88. Find an Old Friend and Call Her
89. Take a Bible Correspondence Course
90. Enroll in a Class
91. Teach Yourself to Use Your Personal Computer
92. Help Children with Homework
93. Explore Your Globe
94. Collect and Label Leaves
95. Study Bugs and Plants with a Magnifying Glass
96. Start a New Hobby

✳ ✳ ✳

84. LISTEN TO MUSIC

Christian music is a good way to fill our minds with whatever is true, pure, and lovely. The Holy Spirit can use godly words to encourage, admonish, or strengthen us.

151

Make a special event out of listening to a new album. Give it your full attention, as you would at a concert. Dim the lights; make a fire in the fireplace; light some candles; pull up a comforter, and snuggle up close.

* * * * *

85. LISTEN TO INSPIRING MESSAGES

Use your Christian radio program guide to find out when your favorite speakers can be heard and plan to listen to them the way you would plan to watch a TV program. You can also check out tapes from your church library. (If your church doesn't have a tape library, consider organizing one.) Listen while you work in the kitchen or the garage. Listen while you drive in the car or do yardwork; or while you stretch out on the couch.

* * * * *

86. LISTEN TO OTHERS

Invite a friend, child, spouse, parent, or neighbor to take a "talk-walk." As you walk around the neighborhood, consider yourself a student of your loved one. How well do you really know him/her? Here are some question/conversation possibilities:

What is your favorite flavor of ice cream?

What is the "funnest" thing you have ever done?

Who are your best friends? What do you like about them?

What is the best and worst thing that happened to you this week?

What is the best thing we have ever done together?

What would you like your life to be like in ten years?

What would you do with $5000?
Is there anything you really regret?
What do you think you are pretty good at doing?

* * * * *

87. RESEARCH YOUR FAMILY TREE

Begin with a trip to the library for a
"How to trace your genealogy" book.
Start corresponding with near and
distant relatives and share your
findings. Use maps to identify the
places your ancestors lived. Check
out some books on the ethnic background and culture of your family.
Find out when the gospel was first
brought to your national group. Are
there many believers in that culture now? Remember that we can
all claim Noah and his wife as our ancestors, and in the pre-
Flood world, Adam and Eve, also.

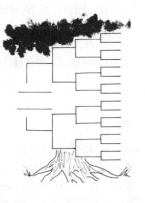

* * * * *

88. FIND AN OLD FRIEND AND CALL HER

Have you ever wondered what happened to your old childhood friend? You used to have great times together, but you
haven't heard from her in years. Call any mutual friends for
information on her whereabouts, then call her. Find out
what she is up to; reminisce about old times. She will be
honored by your detective work, and you will have a great
time sharing each other's news.

* * * * *

89. TAKE A BIBLE CORRESPONDENCE COURSE

Replace the light of the TV with the light of the Word. Enroll in a correspondence course offered by a radio ministry or a Bible Institute. Moody Bible Institute's Correspondence School has a wide offering of courses from New Testament studies to Exposing Cults to Keys to Happy Family Living. Write for their catalog at:

Moody Correspondence School
820 N. LaSalle
Chicago, IL 60610

* * * * *

90. ENROLL IN A CLASS

Look through the local college catalog for classes that will improve your abilities or that just sound interesting. You can audit the class or take it for credit whether you are interested in American history, gourmet cooking, or calligraphy. Many arts or crafts classes are offered by the city parks and recreation service, or consider forming a crafts group of friends. Taking classes is a natural avenue for meeting unchurched friends with whom you can share Christ.

* * * * *

91. TEACH YOURSELF TO USE YOUR PERSONAL COMPUTER

Those who have personal computers may not be using them to their full capabilities simply because they have never

154

taken the time to sit down with the
manual and work through it step-by-
step. Maybe this is the right TV alternative for you. As you gain confidence
and speed, your P.C. will become a
trusted resource which you use regu-
larly and confidently.

* * * * *

92. HELP CHILDREN WITH HOMEWORK

Parents as well as older siblings have a wonderful opportunity to review academics while helping the younger ones in
the family. Whatever the homework is, be available to
answer questions and explain new concepts. Your children
will gain confidence in their ability and in your interest in
them.

* * * * *

93. EXPLORE YOUR GLOBE

Improve your knowledge of geography as you study a globe or map. How
many countries are in the world? How
many can you name? What islands
are in the Pacific Ocean? How many
oceans, continents, and seas can you
name? Have fun imagining faraway
places as you explore your globe. Consult an encyclopedia; use the Global

Pursuit game by the National Geographic Society, or study
a copy of P.J. Johnstones's excellent book, *Operation World*
for interesting facts.

* * * * *

94. COLLECT AND LABEL LEAVES
You will need:
* a book about leaves from the library
* 3 x 5 cards
* pen
* book for pressing leaves between pages
* photo album

As you collect the leaves, look them up in your book and write their names on a 3 x 5 card. Place the leaf and the card together in between the pages of a book. You may even want to make crayon rubbings on paper of the tree bark, as well as a photograph of each tree to include in your leaf album. At home, place your book of fresh leaves under a stack of heavy books. In a few days, take the leaves out of the book and place them in the photo album along with their names and any other information you have gathered. Keep your leaf book out on the coffee table for others to admire and learn from. Use it to quiz your own knowledge. The same idea can be used with a rock collection. Identifying and displaying rocks as well as keeping a journal of where you found them can be a lifelong interest.

* * * * *

95. STUDY BUGS AND PLANTS WITH A MAGNIFYING GLASS
Explore the tiny wonders in your own backyard with the aid of a magnifying glass. Answer these questions:

How many legs does a lady bug have? Do any of your plants have hairy surfaces? What does pollen look like? How does a spider spin its web?Are there any bugs in the bark of your tree? Are there bugs on your roses? What kinds? What is crawling around in your grass?

* * * * *

96. START A NEW HOBBY

You can study and collect stamps, coins, model airplanes, or almost anything else as a family activity, one that will interest everyone. Or try one just for your own enjoyment.

I want to start developing a hobby or collection of _____

_____.

* * * * *

PART EIGHT: THINGS TO MAKE

Young and old alike enjoy the tangible benefits of making things. Activities in this part are not only entertaining but productive as well. Additionally, several of the items suggested will make heartwarming gifts for those you care about.

THINGS TO MAKE - at a glance

 97. Make a Coupon Book
 98. Create a Birthday Card
 99. Make a Memory Album
100. Make a "Day in the Life of. . ." Book
101. Make Christmas Presents
102. Make Doll Clothes
103. Redecorate the Kitchen Bulletin Board
104. Paint a Picture
105. Make Patio Furniture
106. Make Shelves for Your Closets
107. Revive Old Furniture with a Fresh Coat of Paint
108. Make an Educational Board Game
109. Make Puppets
110. Create a Papier-Mache Masterpiece
111. Make Bookmarks
112. Make Paperweights
113. Make Something Fun from a Big Cardboard Box
114. Make a Bird Feeder

* * *

97. MAKE A COUPON BOOK
Think of some things that you can do with or for a friend or family member. Ask yourself, "What help do they need?"

Or, "What would they think is fun?" Give them these gifts of your time in coupon form. Four coupons from an adult to a child might read:

1. This coupon is redeemable for two games of "Chutes and Ladders" with me.
2. The bearer of this coupon is entitled to lick both beaters used to mix cake batter.
3. This coupon is redeemable for two hours of roller skating with a friend this Saturday night.
4. The bearer of this coupon is entitled to a Sunday afternoon at the park with their family and a guest.

A child might give a parent services such as mowing the lawn, washing the car, dishwashing for a week, or organizing the junk drawer. Adults can give adults coupons for babysitting, a pizza delivery, a batch of their favorite homemade cookies, or a date of their choice on Friday night. Cut a piece of 8-1/2 x 11 paper into four strips. Write and maybe illustrate each coupon. Add a cover and a back to your coupon book; then, staple them together at one end and you have a great gift!

I want to make a coupon book(s) for: _____

_____

_____

* * * * *

98. CREATE A BIRTHDAY CARD

Make a birthday really special for someone by taking the time now to design your very own card. Write an honest verse about why you appreciate them. Using high-quality paper, design the cover. Consider stenciling a design, or attach a photograph of the birthday person. Try out your painting or drawing skills for the cover, or cut pieces from used store cards to be combined on your new, original one.

* * * * *

99. MAKE A MEMORY ALBUM

This can be a precious gift for an anniversary or a special birthday. Begin at least two months in advance by secretly asking friends of the honored person to write a short note expressing some trait they appreciate or some experience they remember about the special person. Ask them to include a photograph of themselves for the album. Display the letters and the photographs in a photo album with magnetic pages. (Remember that letters can be reduced on a photocopier if needed to enable you to fit them in the album.) Spruce up each page with flowers or other designs cut from old greeting cards. Use photographs of the honored person (or couple) to illustrate a brief history of their life (lives) to this point. Include captions with dates and important details. Your special person will treasure their memory album for years to come.

* * * * *

100. MAKE "A DAY IN THE LIFE OF. . ." BOOK

Take pictures of your child, spouse, or friend brushing their teeth in the morning, making their bed, playing in the yard, reading, running with friends, or whatever other things they do in a day. Combine the photos in a book to create a story, complete with a narrative you have written. This will make a fun book to read again and again.

160

101. MAKE CHRISTMAS PRESENTS

Anyone who has crammed on December 23rd to finish Christmas presents knows that April is not too soon to begin making Christmas gifts. Brainstorm for ideas by thumbing through hobby and craft magazines. Once you have made your list and gathered your supplies, you will be set until December with productive ideas.

Gift ideas to make:

Person	Gift
_____	_____
_____	_____
_____	_____
_____	_____

* * * * *

102. MAKE DOLL CLOTHES

Every little girl is delighted by homemade doll clothes. Lay the doll, or an old doll dress on a piece of paper and trace it to design your own patterns. Choose a fabric and style that is similar to the little girl's own dress. This is a great way to make use of your fabric scraps. Even if you don't have any little girls of your own, hospital children's wards, your church benevolent collection, an orphanage, or homeless children's shelters will be thrilled with your gifts.

* * * * *

103. REDECORATE THE KITCHEN BULLETIN BOARD

Replace the outdated phone numbers and messages on your bulletin board with some seasonal borders or decorations. Be sure to attach a fresh pad of paper and a pencil.

*****

104. PAINT A PICTURE

Could it be that you are one of those people who says, "I'd like to try my hand at oil or water color painting someday?" Well, maybe now is your big moment. Gather your equipment and plunge in. Allow yourself plenty of practice and enjoy learning through doing. Store your painting supplies in one convenient location where they will be easy to get to the next time you are in the mood to paint.

*****

105. MAKE PATIO FURNITURE

Check out your favorite do-it-yourself book for instructions on cutting and assembling a planter for the patio or maybe a porch swing or lawn chair. Your creative efforts will be admired and appreciated by many.

*****

106. MAKE SHELVES FOR YOUR CLOSETS

Do you feel like you are running out of storage space? The

answer may be simpler than moving to a new house. Chances are, you could use your closet space much more efficiently by adding a few more shelves. Once you have taken measurements and secured your supplies, the project can probably be completed in one evening.

* * * * *

107. REVIVE OLD FURNITURE WITH A FRESH COAT OF PAINT

Dig out an old stool, chair, table, or bookcase. You might find just the thing one of your rooms has been needing. Give the item a fresh coat of paint, and then enhance it further with a stencil design or a contrasting color of trim. You may create a cherished new item for your home.

* * * * *

108. MAKE AN EDUCATIONAL BOARD GAME

Create an original, homemade board game for the kids or the whole family. You can make reviewing academics as much fun as developing strategy.

Board
A game board can be made from a piece of poster board, a rectangle cut from a cardboard box, or a file folder opened up flat. For the most basic game, begin by writing "start" in an upper corner and make a trail of boxes which wind around the board, finally reaching the bottom, opposite corner of the board, and the word "end." Intersperse the trail with phrases such as: "Sorry, move back two spaces" or "Hurray! Move ahead four spaces."

Cards

Use 3 x 5 cards for question cards. On each one, write a question. If you are practicing geography, ask questions about capitols and country locations. If you need to review math facts, use flash cards for your question cards. Just about any subject can adapt to this format including history, literature, music, reading, and science. Write answers on the backs of the cards or on an answer sheet.

Place markers and dice

Round, plastic milk carton caps with colored stickers or buttons make good place markers. Use a die or a game spinner to determine how many spaces each player may advance.

How to play

Two or more players begin by choosing who will go first. The first player places his cap on "start" and selects a question card without looking at the back of the card. If he answers correctly, he may roll the die and advance accordingly. Each player in turn selects a card, tosses, and moves. The winner is the first player to reach the end box.

Variation

Pick a theme for your game board such as "Space Flight" or "Marathon" or "The Circus" and decorate the board with pictures that reflect the theme. Change "start" and "end" to "take-off" and "touch down", "starting line" and "finish line," or "ticket booth" and "grand finale." Be creative and imaginative, capturing your family's specific interests.

* * * * *

109. MAKE PUPPETS

Popsicle Stick Puppets
Cut people and animals out of coloring books or Sunday School papers. Glue them to a heavier paper backing and then attach a popsicle stick with glue or tape.

Sock Puppets
Put your hand in the foot of an old sock. Determine where the mouth, eyes, and hair should be. Cut these out of felt and yarn to be glued or stitched in place. These puppets are some of the best because they can move their mouths.

Finger Puppets
Draw a face on your fingertip with a pen. Make clothes for your puppet by taping tissue paper skirts and pants onto your fingers.

Glove Puppets
Glue felt pieces and pompons onto the fingers of a pair of white gloves. Add facial features and you have ten wiggly puppets ready to tell a story.

Paperbag Puppets
Cut faces out of construction paper and glue to a small, lunch-size paperbag. The folded bottom of the bag forms the mouth.

Paper Plate Puppets
Staple two paper plates together around the edges leaving an opening at the bottom for inserting your hand. Draw a face on the front. Attach ribbon or yarn for hair.

Now, make up a story and act it out for the family using your very own puppets. A theater stage can be made from cardboard boxes. Let imaginations really soar for this performance.

*****

110. CREATE A PAPIER-MACHE MASTERPIECE
Papier-mache crafts are created by wrapping pasty strips of newspaper around boxes and wads of newspaper to make a new shape or figure. Begin by making your own paste from this recipe:

Flour and water paste
Homemade paste can be made by mixing together 1 cup of flour with 1/2 cup of cold water until smooth. Pour 2 cups of boiling water over the mixture and stir until it becomes transparent looking. Dilute with water until it is the consistency of heavy cream.

Tear or cut newspaper into 1/2 inch strips. Pour some of the paste into a shallow dish. Pull the strips of paper through the paste and wrap around the form. (Wire forms of many designs can be purchased at craft stores.) Continue layering the strips placing each new layer at right angles to the layer beneath it until it is about 1/4 inch thick. Let dry, sand, and paint.

To make a piggy bank
Tape or tie stubby cardboard toilet paper rolls as legs to a cylindrical oatmeal box which is the pig's body. A ball of waded newpaper with a stubby snout can be the pig's head. When the pasty strips are attached, ears and other details

can be added. Cut a slit in the top for the coins to slip through and a hole in the bottom for a cork stopper. When the paste is completely dry, sand down the rough spots, give the bank a coat of paint and a happy pig face, then cover it with a shiny coat of shellac.

Use this same method to make masks, helmets, contour maps, etc.

* * * * *

111. MAKE BOOKMARKS

Cut a piece of felt into a strip 10 inches long and two inches wide. Cut 1-1/2 inch slits into the top and bottom of the felt strip to create a fringed appearance. Decorate the bookmark by gluing or sewing on strips of embroidered ribbon or lace. Children can make these for special friends or family members and use as birthday or Christmas gifts.

* * * * *

112. MAKE PAPERWEIGHTS

Gather smooth stones for your paperweights. Paint them to look like ladybugs or faces, or paint pretty flowers and designs on the stones. Cover with a coat of shellac. These also make creative and personal gifts for children to give.

* * * * *

113. MAKE SOMETHING FUN FROM A BIG CARD-BOARD BOX

With a good imagination, some paint, crayons, and maybe even scissors, you can transform a big box into a motor home or a boat. You can make it a house or an airplane. Draw in wheels and gauges. Cut out windows and doors. If you make a boat, bounce around in it a bit as you pretend you are sailing through turbulent waters. If you make a house, pillows and blankets can become chairs and carpets. Tape paper towel curtains at the windows and invite friends over. Serve cookies!

* * * * *

114. MAKE A BIRD FEEDER

Pie tin bird feeder

You will need a disposable pie tin and four pipe cleaners. Poke four evenly spaced holes around the rim of the tin. Insert the pipe cleaners and bend the ends to secure. Twist the pipe cleaners together at the top where they meet and attach a loop. Hang it from a tree branch. Fill with birdseed or crumbs and have fun bird watching.

Pine cone feeder

Attach a loop to the top of a large pine cone. Smear peanut butter all over the pine cone being sure to fill all the undersides. Now roll it in birdseed and hang it in a tree for some hungry little birds.

* * * * *

PART NINE: OUTREACH ACTIVITIES

"Love one another" is one of Scripture's primary commands, yet we sometimes have difficulty expressing that love. A plaque I admired in the home of a missionary read: "Love is an action word." How true! We need action if the world is ever going to believe that God loves them. Reach out to those in need physically, emotionally, and spiritually.

As we free ourselves from the TV, we can look outside our four walls for ways to show Christ's love. Giving your time to another person is about the number one best way we can say "I love you." In fact, getting involved in people's lives is one of the best ways to insure that we will *continue* to watch less TV. Look through the alternatives that follow and find something that can help you put *your* love into action.

OUTREACH ACTIVITIES - at a glance

115. Do Repairs for Someone in Need
116. Walk and Talk with Your Neighbors
117. Visit an Elderly Person
118. Host a Neighborhood Children's Carnival
119. Plan a Neighborhood Ice Cream Social
120. Host a Good News Club
121. Host a Girl's Tea Party
122. Host a Boy's Root Beer Float Party
123. Walk Someone's Pet
124. Take Care of a Homebound Person for a Day
125. Volunteer at a Non-Profit Ministry
126. Help Out in the Children's Ministries at Your Church
127. Organize a Church Work Day
128. Prepare and Serve a Meal at a Rescue Mission
129. Volunteer

115. DO REPAIRS FOR SOMEONE IN NEED

Is there something that you can do for a neighbor, friend, or church member? Think of those you know who are sick, elderly, handicapped, just had a baby, experienced a recent death or divorce in the family, or are facing some other crisis. Don't forget the single moms who are trying to meet the responsibilities of both mom and dad in the home. If you can't think of anyone in need, call your church office for some suggestions.

Then, give your friend a call and tell them that you are planning on doing a few chores or repairs for them. Where would they like you to start? If it is a big project like replacing the roof, invite other friends to join the team. Here are just a few suggestions for ways you can help:

- wash their car
- mend clothes
- change the car oil
- repair a leaky faucet
- wash second story windows
- mow the lawn
- shovel the walk
- rake the leaves
- weed the garden
- clean out the gutters
- baby sit
- run errands

* * * * *

116. WALK AND TALK WITH NEIGHBORS

Take a walk with the intention of stopping to talk to any of your neighbors you happen to see outside along the way. Or, begin by inviting a neighbor to join you for a walk together. Eventually, you may get the privilege of sharing Jesus with them, also.

117. VISIT AN ELDERLY PERSON

Some of the loneliest people in the world may live on your street or attend your church. Elderly people often know the pain of seeing their best friends or spouses die and are starving for companionship, yet are striving not to be a "burden" to anyone. Visiting the elderly is for kids and adults alike. If you don't already know some older person you can visit, call the local nursing home and ask them to direct you to someone who would appreciate regular visits. You can:

* Spend an hour or two just listening and talking
* Invite them over for dinner.
* Take them out for a ride in the car.
* Send them notes in the mail.
* Read to them.

* * * * *

118. HOST A NEIGHBORHOOD CHILDREN'S CARNIVAL

This event will be great fun for the children on your block as well as provide a great opportunity for the neighbors to work together and get to know each other. Ask each family or pairs of families to prepare a carnival booth or table with some game or activity for attendees to enjoy such as:
* ring-toss game
* throw the frisbee through the hoop
* make basketball free-throws
* a cake walk
* a dunk tank

- a fishing booth where simple prizes are hooked
- guess how many jelly beans are in the jar
- face painting
- soap box derby
- a popcorn and lemonade stand

Encourage children to do a large part of the planning and running of the carnival day.

Hold your carnival in a large backyard, or arrange for the police to barricade the street for the evening so that you can set up in the street. Have a potluck dinner and then show a Christian movie outdoors after dark. Maybe you can borrow your church's projector and screen.

* * * * *

119. PLAN A NEIGHBORHOOD ICE CREAM SOCIAL

A fairly easy neighborhood party to plan is an ice cream social. Schedule it for a Sunday afternoon and ask everyone to bring a half gallon of ice cream, a scoop, and at least one topping to share. You arrange for the tables and chairs, paper bowls, cups, and plastic spoons. Provide water or some beverage for thirsty ice cream  eaters. You might ask the police to set up barricades for the event and have your party right in the middle of the street. Plan some simple get-acquainted games and relay races to get people interacting. Use the party to strengthen friendships and as a jumping-off point for inviting neighbors into your home at some other time.

* * * * *

120. HOST A GOOD NEWS CLUB

If you have ever wondered how to reach the kids on your block with the good news of God's love, a Good News Club may be just the thing for you. Child Evangelism Fellowship, an interdenominational, non-profit organization, has devised a series of successful programs geared to attract school-aged children to your home for club meetings and lessons from the Bible. A trained teacher will come into your home to lead songs, puppet skits, Bible lessons, verse memorization, games, and handcrafts.

They offer several different programs: Nine-month or Six-Week Good News Clubs meet after school for one hour each week. The Five-Day Club meets outside on a lawn everyday for five days during the summer. Club attenders are then enrolled in a correspondence course as part of their follow-up. The Party Club is designed for you to host evangelistic parties once a month, several times a year or only once a year. Find out how to get involved in one of the programs by calling your area CEF representative. Find "Child Evangelism Fellowship" in the white pages of your phone book or contact the organization's headquarters, at:

Child Evangelism Fellowship
P.O. Box 348
Warrenton, MO 63383
Phone: (314) 456-4321

* * * * *

121. HOST A GIRL'S TEA PARTY

Girls will have fun planning and decorating for this special little event all by themselves. They can send out homemade invitations. Decorate the table and arrange for snacks. You can serve tea or any other beverage children enjoy but be sure to pour it from a teapot into pretty teacups. Remember to include some special treat which you don't usually serve such as mints or cut-out cookies which the children decorate. The young hostess can learn the joy of sharing the good news with her guests by reading a gospel story from a book or telling it with flannelgraph figures. Another possibility is to show a Christian children's video after tea time. Check the video reviews in this book for suggestions.

* * * * *

122. HOST A BOY'S ROOT BEER FLOAT PARTY

Boys can prepare for an afternoon root beer float party by inviting a few friends and asking them to bring their favorite or most unique mug or glass. Serve the floats in the boys' mugs. After the snack, the boy-host can share the gospel with his friends by reading an exciting Christian story or showing a Christian video. You may want to continue your meetings together and form a kind of "club" that gets together to snack, play, and read a continued boy's Christian adventure book.

* * * * *

123. WALK SOMEONE'S PET

Kids can offer to help a neighbor, have fun, and get exercise all at the same time. Take their dog for a run around the block. They are likely to make two new friends.

* * * * *

124. TAKE CARE OF A HOMEBOUND PERSON FOR A DAY

Give the parent of a handicapped child or the child of an aging parent a day out of the house. Arrange to learn their responsibilities and allow them some much needed time away from their constant responsibilities of caregiving. You will gain new understanding and compassion for the handicapped as well as their caretakers. Contact your church or local health agency to put you in touch with someone who would appreciate your help.

* * * * *

125. VOLUNTEER AT A NON-PROFIT MINISTRY

Make a list of the ministries in your area. Select the ones which you are most interested in helping and give them a call. Tell them that you would like to volunteer two hours or so a week. Ask if there is anything that they would like you to do. You may find yourself answering phones, counseling unwed mothers, stuffing envelopes for mailings, sweeping floors, typing, filing, or visiting patients. The possibilities are endless. You may not be able to contribute financially, but you will be contributing a great deal when you give dependably of your time.

* * * * *

126. HELP OUT IN THE CHILDREN'S MINISTRIES AT YOUR CHURCH

It's surprising that there appears to be a shortage of children's workers in our churches. There are plenty of opportunities for everyone. Let your church know that you would like to help out. You are needed, whether your abilities lie in storytelling, song leading, piano playing, game leading, listening, praying, being a friend, driving cars, preparing food, or organizing records.

*****

127. ORGANIZE A CHURCH WORK DAY

There are many projects around the church that never seem to get accomplished until members join forces. Many hands make light work as well as good company. You can be a real blessing to your pastor and church staff by coordinating a work day for your church. Get or make a list of projects which need to be accomplished. Gather the necessary supplies. Set a date for your work day and promote it well. As workers arrive, assign them tasks in small groups and watch how quickly the jobs can be accomplished. Arrange for a lunch to be provided for workers, or have everyone bring a sack lunch.

*****

128. PREPARE AND SERVE A MEAL AT A RESCUE MISSION

Volunteer to help in the kitchen of a rescue mission or ministry which provides meals for the homeless. Your church, Sunday School class, or family can take responsibility for providing and serving one meal a month. Put your Christianity into action; call up your local mission or shelter to learn what you can do.

Quick reference information:

Shelter name _____

Address _____ Phone #: _____

Needs _____

* * * * *

129. VOLUNTEER

Some Christians never get out of their Christian circle of friends to rub shoulders with people who need to know Christ. One way to solve this problem may be to volunteer your services at the public library, a local hospital, or in your child's public school classroom. You will help your community as well as make new friends with whom you may someday share Christ.

* * * * *

PART TEN: FORGOTTEN WORK PROJECTS

Every household has jobs that tend to be forgotten or neglected. These are usually the maintenance items that only require once a month or even once a year attention. This is a list of projects that are typically forgotten or put off. At the end of the list is space where you can jot down your own personal job reminders. Before you flip on the TV, why not complete some of these projects? You will enjoy a sense of accomplishment when you finish tasks that never seem to get done.

FORGOTTEN WORK PROJECTS - at a glance

130. Organize Your Photos into an Album
131. Clean Out Your Junk Drawer (Junk Room!), Tool Chest, Closet, Attic or Garage
132. Simple Home Repairs
133. Clean Out Your Files
134. Mend Clothes
135. Shine Shoes
136. Wash Windows
137. Clean Out the Gutters
138. Wash the Car
139. Clean the BBQ Grill
140. Wash the Dog
141. Check your own list for more ideas

_____

_____

_____

* * *

130. ORGANIZE YOUR PHOTOS INTO AN ALBUM

Are the photos you have taken piling up? Organizing them into photo albums can be fun. The whole family can be involved as they look at pictures of themselves and re-live happy memories.

*****

131. CLEAN OUT YOUR JUNK DRAWER (JUNK ROOM!), TOOL CHEST, CLOSET, ATTIC, OR GARAGE

Do you have difficulty finding a place to put things? Do you have trouble finding the items you are looking for? If so, it is time to clean out and organize your junky drawers, tool chest, closet, attic, or garage.

A simple project like organizing drawers can be done by declaring a drawer organizing night when everyone in the family works on the same project. On the other hand, a project like cleaning the garage may take the whole family's efforts one Saturday or over several evenings until little by little the whole garage is neat and organized.

*****

132. SIMPLE HOME REPAIRS

Every household has a few irritating maintenance needs that could be easily repaired if someone just took the time to fix them. Begin by making a list of needs throughout the week. The next time you are tempted to watch TV out of habit rather than conscious choice, pull out the list and choose one of the items to repair. You will be glad you did.

*****

133. CLEAN OUT YOUR FILES

Filing cabinets have a way of accumulating vast quantities of paper that are never seen or retrieved again. Periodically, everyone needs to flip through their files to reacquaint themselves with the contents as well as clear out unnecessary items.

*****

134. MEND CLOTHES

Gather together all the clothes that are in need of repair. Sew on buttons, restitch seams, repair pockets, adjust hems, and fix zippers. In a few hours you dispose of a bundle of little annoyances as well as revitalize a whole wardrobe.

*****

135. SHINE SHOES

Instead of running out the door with scuffy looking shoes, take a few minutes now to shine them. The next time you are dashing around getting ready for a dressy occasion, you will thank yourself for a pair of ready-to-go, clean and polished shoes.

*****

136. WASH WINDOWS
Get out the window cleaner and some rags and wash the windows. Maybe you can entice a friend into washing the insides while you clean the outside. The fresh, clear panes will give you a brighter outlook on the world.

* * * * *

137. CLEAN OUT THE GUTTERS
Choose a pleasant day and a sure-footed ladder to clear your gutters of the debris that accumulates through the season.

* * * * *

138. WASH THE CAR
Whether you drive a jalopy or a Corvette, a good clean-up job improves one's pride of ownership. After you have washed, vacuumed, scrubbed, rinsed, dried (and maybe waxed!), treat yourself to a scenic drive with your sweetheart.

* * * * *

139. CLEAN THE BBQ GRILL
This is everybody's least favorite job, so do the family a favor and scrub that grungy BBQ grill. When your mouth is watering for some delicious hamburgers, the last thing you want is to put them on a dirty grill. Get this chore out of the way and you will be ready to enjoy your next Bar-B-Q.

* * * * *

140. WASH THE DOG

With the right attitude, this job can
be a fun one. Suds and rinse Fido
until he looks and smells as lovely as
a rose again. You will appreciate his
fresh appearance and Fido will enjoy
the attention he receives.

* * * * *

141. LIST YOUR OWN WORK PROJECTS

_____

_____

_____

_____

_____

* * * * *

PART ELEVEN: PLAN-AHEAD ACTIVITIES

Activities in this category take a little time and planning. Some people find the planning as much fun as the doing. At any rate, pull out a piece of paper and jot down your creative ideas for these plan-ahead activities.

PLAN AHEAD ACTIVITIES - at a glance

142. Secretly Plan an Extra-Special Date
143. Plan a Fun Date With One of Your Children
144. Plan a Vacation
145. Design and Build a Tree House
146. Organize a Neighborhood Craft or Garage Sale
147. Draft a Will
148. Develop a Family Budget
149. Plan a Treasure Hunt
150. A Special Idea Just Your Own

142. SECRETLY PLAN AN EXTRA-SPECIAL DATE

What do you think your loved one would really enjoy on this extra-special date? Does he like baseball? Plan on taking in a game on your special day. Does she enjoy art galleries? Include it as part of your date. Maybe you will want to arrange for some unusual transportation like a horse and buggy or a limousine. Consider making it a whole day event beginning with breakfast, then outdoor fun, and dinner at an elegant restaurant. Married couples could consider an overnight stay at a local resort.

For many zany date ideas read *Creative Dating* by Doug Fields and Todd Temple. It's available at Christian and secular bookstores. However you personalize your date, make it an extra-special evening he or she will cherish for a long time.

* * * * *

143. PLAN A FUN DATE WITH ONE OF YOUR CHILDREN

Kids are delighted in knowing that you planned a special date just for them! Send your child an invitation in the mail for a "Fun-date with Dad (or Mom) on Saturday." The added suspense and anticipation can make the event even more exciting. Maybe you will want to make it a yearly event celebrated on each child's half-birthday (a day six months from their real birthday). Cater the date's activities to things the child enjoys. Start conversations that give them an opportunity to express their feelings. Be vulnerable yourself and share some of your feelings. Express some of the qualities that you appreciate about your son or daughter.

* * * * *

144. PLAN A VACATION

Give your vacation plans some time to form this year by planning far in advance of your departure date. Begin by taking an informal survey of what each person in the family thinks makes a good vacation. Jot down the different ideas and try to incorporate at least some of their desires into your plans. Determine your dates and budget. Check out a travel book from the library such as Reader's Digest's *America From The Road* or *Off the Beaten Path*. Be careful not to plan so much travel that your time is spent solely on interstate highways. Plan where you will stay, what sights you want to include, how much money you can allot to each day's activities, and how you can make the in-car travel time fun and relaxing. Highlight the sights you will be seeing on the map. Include older children by allowing one teenager to be in charge of the daily finances and another to be in charge of in-car entertainment and morale. Have fun making your plans but determine to be flexible when things don't go the way you planned.

* * * * *

145. DESIGN AND BUILD A TREE HOUSE

It may be that you have always wanted to build a treehouse, but you have never gotten around to it. Make your evenings more entertaining by pulling out paper and pencil and coming up with a realistic sketch and plan for what you want the treehouse or playhouse to be. Get a book from the library or hardware store that suggests building plans. Include your children as you purchase supplies and begin building. Even if they are too young to hammer, they will be thrilled to hand you the nails and help with painting and sanding. When a child feels like she was an important part of a project, she will have more appreciation and respect for it.

* * * * *

185

146. ORGANIZE A NEIGHBORHOOD CRAFT OR GARAGE SALE

Set a date and a location and invite your friends and neighbors to contribute their homemade or used items. Come up with a plan for managing the money, manning the sale, advertising, displaying items, and redistributing profits and unsold goods. Call a casual meeting in your home to inform your neighbors and ask for their help with specific tasks. Many problems can be avoided if everyone is properly informed. The sale may be just the vehicle you've been looking for to develop friendships and make a little extra money as well.

* * * * *

147. DRAFT A WILL

If drafting a will is something which you have been putting off, now is the time to plan for the future. Seek out a lawyer and take care of it. For most families it is a simple matter which can be settled with one visit to the lawyer's office. Prior to your meeting with the lawyer you will want to consider these items:

* Who will you name as the preferred guardian for your children should you both die?
* Who will be the estate executor/executrix?
* Do you want to establish a Living Will?
* Do you want specific possessions to be willed to specific people?
* Will you leave a portion of your estate to a church or ministry?
* Would a Living Trust be a better way to distribute your estate?

Involve adolescent and older children in these discussions

so it is less threatening or uncertain for them. Consider their wishes for guardian, etc., too.

* * * * *

148. DEVELOP A FAMILY BUDGET

Many families find themselves in a position where their expenses exceed their income. If you are one of those families, you need to sit down with pencil and paper and determine where your money is going as well as where you would like it to go. Set up a realistic budget and learn to live within your means again. For materials dealing with financial management, we suggest you call or write to the organization founded by author and radio speaker, Larry Burkett: Christian Financial Concepts

        601 Broad St. S.E.
        Gainsville, GA 30501
        Phone: 1-800-722-1976

Another excellent book on the subject is Amy Ross Mumford's *It Only Hurts Between Paydays*. By involving all family members, the children will begin to understand financial responsibility as well as why they can't have everything they want. Help them become responsible stewards, too.

* * * * *

149. PLAN A TREASURE HUNT

Write and hide clues that will lead your treasure-seeking guests to one hidden clue after another until at last they are given the final clue which should lead them to the "treasure." Your treasure can be a box of foil-covered chocolates, a book, or a Christian video. To play in teams, make up two equally difficult sets of clues that lead to the same treasure. The team that reaches the treasure first is the winner.

* * * * *

150. A SPECIAL IDEA JUST YOUR OWN
Write out the basics of a special event custom-made for you or your family.

CHAPTER TWELVE

THE VIDEOGUIDE:
What's Available, What's Good—and Where to Find It!

The owner of a flourishing video rental store found himself in quite a quandary. Having recently dedicated his life to Christ, he was increasingly troubled by the type of videos that he was renting and selling. His shelves were laden with more than 1600 different titles, but among them were very few that he felt good about. In fact, he stated that "if the Lord Jesus walked into (my) video shop and asked for a decent video, out of the 1600 on the shelves I could probably only scrape together 12!"[1] That realization hounded his conscience. Even though he had invested thousands of hours in the hard work of building his business, and even though he would face a large financial setback, he knew that he had to get out of the business. And he has.

Most secular video movies portray an even coarser degree of the same types of offensive behavior that we try to avoid on TV. It's the same old story: Free to choose but nothing to choose from. Author and radio talk-show host Bob Larson categorized the films from a recent season into their rating groups. This is what he found: "Looking at all the films during a recent film season, these are the categories by percentage: R and X - 56%, PG and PG-13 - 40%. In other words, 96% of all films were unsuitable for family viewing."[2]

Most Christian viewers know that these "Hollywood" products are not appropriate for their consumption, but they consume them anyway. Why? Because all of us want to sit back and be entertained from time to time. Sadly, the majority of us simply are not aware of the many excellent videos available with a Christian worldview. Christian movies are not reviewed by Siskel and Ebert. Nor are Christian movie producers endowed with the millions of production and advertising dollars necessary to buy their way into the secular theatrical market. Therefore, their existence largely remains a "secret," except to the small percentage of innovative pastors who exhibit Christian films to their congregations, and to the families who happen to live in the vicinity of a Christian bookstore that offers a selection of these movies on video for in-home use.

A videocassette player can be an excellent tool in the hands of discerning and well-informed family leaders. Adults can provide their families with videos from the vastly underutilized pool of excellent Christian dramas, documentaries, children's shows, and "how-to" tapes which are now available.

We have developed this VIDEOGUIDE section after personally evaluating every tape listed. This section has been carefully compiled to help you to know what tapes are truly worthy of the investment of your time and money, and what tapes to avoid.

It should be clearly understood that the VIDEOGUIDE does not attempt to summarize or evaluate traditional/secular movies. This job is already being handled by several authors and organizations. A list of books and newsletters published for this purpose is shown in the "Footnote" section of this chapter[3]. In contrast to these materials, our VIDEOGUIDE is designed to showcase productions that have been written and produced from an explicitly Christian world view.

SPONSOR A CHURCH VIDEO LIBRARY

With the rapidly increasing popularity of VCRs, members of your church can have a media ministry right in their own homes. Videos with a Christian message can be loaned to non-Christian neighbors—or better yet, non-Christian friends can be invited over to share fellowship and the message of the program. Whether with serious drama or children's tapes after school, carefully selected videos can provide a natural, friendly way to communicate our faith.

Encourage your pastor or church librarian to contact the distributor at the phone number listed with each VIDEO-GUIDE evaluation, or to call your local Christian bookstore or film library. Also be *sure* that you clearly understand whether the tapes that you are purchasing are cleared for "Public-Use" (i.e., church services, Sunday School classes, VBS programs, youth camps, etc.) or if they include only the standard private "Home-Use" rights. Like their Hollywood counterparts, most high-quality Christian tapes are restricted to use in private residences only and cannot legally be exhibited to church groups no matter what the video store clerk may tell you! Adherence to this limitation by the Christian community is vital to the continued existence of

Christian film producers. A higher, "Public-Use" fee must often be paid to the producer/distributor in order to be able to exhibit a tape in public legally, even to a church group. Only the copyright *owner* can legally grant Public-Use showings of copyrighted tapes. Get such authorization in writing.

If it is included, the additional "Public-Use" authorization will usually be printed on the back of the video package and on the label of the videocassette itself. Some producers offer the convenience of being able to "upgrade" Home-Use tapes for showing to church groups for an additional fee. An upgraded license certificate is sent to the video owner as proof that he has paid the necessary amount for specific Public-Use rights, too.

Be scrupulously honest! If the tape that you want to show to your church group doesn't specifically authorize public showings—*in writing*—DON'T SHOW IT UNTIL FIRST CHECKING WITH THE DISTRIBUTOR WHOSE NAME IS PRINTED ON THE VIDEO PACKAGE! Every copyrighted video is automatically restricted by law from any public showings, no matter how small the audience, unless additional rights are specifically granted.

Christians must come to grips with the fact that they are not immune from copyright law. While the gospel message is free, putting it into an entertaining format by using paid scriptwriters, directors, actors, soundmen, sets, etc., is not. This can cost hundreds of thousands of dollars. And most Christian producers operate on an extremely tight budget. It is unsettling to hear pastors and lay leaders say that copyright infringement is necessary or justified if its purpose is to advance the work of the Lord. God's command is clear, "Do not steal" (Romans 13:8-9). "For we are taking pains to do what is right, not only in the eyes of the Lord but also in the eyes of men" (II Corinthians 8:21). Illegally exhibiting videotapes in public-use settings pulls the financial rug right out from under the producer's feet!

As you read through the following video reviews, you will see that most of the higher quality productions are also available in the 16mm film format. If the film is a good one, many churches buy the video version for the church library. Consult our VIDEOGUIDE or your local Christian video store to put together the best possible video lending library that your church budget will allow.

HOW THE VIDEOGUIDE WORKS

The following VIDEOGUIDE provides you with a standard for considering the relative value of tapes before buying or renting them. Even those individuals who have seen only one or two Christian films can gauge the quality of other shows listed by comparing the rating of the titles they have seen to those which they have not.

RATING - (One to five stars) A rating of up to 5 stars has been assigned to each production in each of two areas: "Content" and "Technical Quality." The definition of each rating is generally described as follows:

★★★★★ 5 star . . . EXCELLENT!
★★★★ 4 star . . . VERY GOOD
★★★ 3 star . . . GOOD
★★ 2 star . . . FAIR
★ 1 star . . . POOR

CONTENT - While technical aspects are often the most noticeably excellent or horrible parts of a program, they are not the most important. Secular producers have created thousands of *technically* excellent movies. However, only a handful have contained a message that was truly worthy of the viewers' time. In assigning a rating of one to five stars

193

for each production, we have posed and answered the following questions:

1. Does the production claim to be based on a biblical record or story? If so, how true to the scriptural account is it? If it adds to the story, are its additions logical and probable in light of other historical, scientific and/or cultural knowledge that is available? Does the production overlook or gloss over certain aspects of the story? If so, are the missing items vital to communicating important ideas or events?

2. Is this production appropriate for viewing by the intended age group?

3. How well is the script written and how well is the intended message communicated? Does the audience remain attentive and have a good understanding of the message or is it left wondering just what the intended message really was?

Of overriding concern in this category, especially when evaluating children's videos, is question #1. If the production claims to be based on a Bible story, but presents material not in keeping with scriptural, historical, and cultural knowledge, the production could not score highly in this category even if the script and entertainment aspects were excellent.

TECHNICAL QUALITY - Again using the one-to-five-star rating system, each show was evaluated with regard to its technical considerations. Those considerations include such aspects as the quality of editing, cinematography, animation, soundtrack, acting, etc.

194

PRIMARY AUDIENCE - The age group listed is that which we feel would derive the greatest benefit from viewing the show. This is not to infer that a wider age range would not also enjoy the production.

DISTRIBUTOR - The large majority of tapes listed in the VIDEOGUIDE can be purchased at (or at least ordered through) any Christian film library[4] or bookstore. In case you do not live near one of these retail outlets, we have listed the name and phone number of the video's actual producer, or a knowledgeable distributor. The company listed should be able to answer any questions you may have about content, usage restrictions, prices, and ordering information.

OUR COMMENTS - The comments that we have included with each video evaluation are designed to cut through the marketing hype sometimes written by those who are trying to sell the tapes. We have tried to give a brief summary of the topic, story, or message communicated, and to provide an honest appraisal of the overall value of each production.

It is our earnest prayer that our evaluations will serve to open your imagination to some of the many uses possible for the broad range of materials now available. Whether you want to trim your waistline by doing aerobic exercises to Christian music, learn about Creation and how dinosaurs fit into the Bible, watch and discuss with your nine-year-old a Bible-based video about sex, learn how to catch Bigmouth Bass, or simply enjoy a bowl of buttery popcorn while relaxing in front of an inspiring family drama, there is a Christian video out there to meet your needs.

When managed with the help of the ideas, forms, and alternatives provided in this book, your TV can become a wonderful tool. You don't have to be married to television!

<u>FOOTNOTES</u>:

[1]Creation Science Foundation Prayer News, April/May 1987, p. 3 (Creation Science Foundation, Queensland, Australia).

[2]Bob Larson, *Larson's Book of Family Issues*, (Wheaton, Illinois; Tyndale House Publishers, Inc., 1986), p. 269.

[3]The following books and newsletters are provided to aid Christians in their decisions about what movies are appropriate for viewing and what ones should be avoided.

- *The Christian Family Guide To Movies And Videos*, 2-volume book set (Wolgemuth & Hyatt Publishers, Brentwood, Tennessee, 1989) by Ted Baehr. Contains hundreds of secular video evaluations. Available from any Christian bookstore.

- *Movieguide*
  Monthly newsletter gives detailed secular movie reviews and lets readers know why they should avoid or attend the movie in question. Gives a biblical perspective of each movie to help develop biblical world view and discernment. $30 per year.
  Contact: Good News Communications
  P.O. Box 9952
  Atlanta, GA 30319
  Phone: 404/237-0326

- *Movieguide*
  This 900 line provides insightful, recorded reviews of your choice of up to five current movies or five recent video releases from a Christian perspective. At the time of printing this book, the cost for the phone call was $ .65 per minute.
  Contact: Good News Communications
  1-900-226-3400

- *Movie Morality Guide* (newsletter)
  Thoroughly describes and rates all popular PG and PG-13 movies according to both entertainment value and content acceptability, published twice monthly. $25 suggested donation.
  Contact: Preview
  1309 Seminole Dr.
  Richardson, TX 75080
  Phone: 214/231-9910

- *Scoreboard Alert*
  Newsletter covering various issues important to evangelicals.
  Includes simple chart listing movies by title with quality and
  content ratings. No summaries or evaluations. $24 per year.
  Contact: National Citizens Action Network
  P.O. Box 10459
  Costa Mesa, CA 92627
  Phone: 714/850-0349

⁴Christian film libraries can be located simply by asking your
pastor for the name of the company from which he rents Christian films, or by calling the headquarters of the following
organization and asking that a complete video catalog be sent to
you from the Christian film library nearest you.

- Christian Visual Media International (CVMI)
  Phone: 303/761-0092

THE VIDEOGUIDE

ACTS: Volume 1

Content: ★★★★1/2
Technical Quality: ★★

PRIMARY AUDIENCE: Ages 5 - Adult
STYLE: Filmation

LENGTH: 38 min.
DISTRIBUTOR: Biblevision
209/825-5645

This video covers Acts 1-7 with integrity and scriptural purity. This series of four tapes combines a word-for-word dramatized version of the NIV translation of the entire book of Acts (including sound effects and music) with pans, zooms, and dissolves of nicely detailed paintings. The illustrated characters are realistic, not cartoonish. An optional "Teacher's Guide and Video Game Book" includes suggested questions to help review every segment and age-graded game ideas to benefit viewers beginning at about 8 years. Well-suited to small group and family Bible studies, especially when used in conjunction with the Teacher's Guide. The visual and verbal panoramic overview of Jewish history by Stephen in Acts 7 (just prior to his stoning) is a highlight of this tape. This series should be in every church video library! It's perfect for dads who need help to initiate or revitalize family devotions.   ©1986

A.D.
(Abridged edition for churches)

Content: ★★
Technical Quality: ★★★★★

PRIMARY AUDIENCE: High School - Adult
STYLE: Drama

LENGTH: 6 hours (on 3 tapes)
DISTRIBUTOR: Gospel Films Video
616/773-3361

This multi-million dollar production was filmed on location in Tunisia, Pompeii, Herculeneum, and Rome. Originally broadcast in the USA as a network mini-series in the mid-1980s. Covers the years from 30-69 A.D. and "focuses on the rising confrontation between the mighty Roman empire, Jewish zealots, and the early Christians. Mixes historic incident, biblical narrative, and the result of years of research into customs and conditions in the first century." Not scripturally precise but close enough that, if used in conjunction with the study guide provided, carefully led discussion times following each of the twelve segments can reveal and correct most of the important misrepresentations. Soap-opera type storyline includes fictional characters, romance, violence, and killing. NOTE: Extremely violent and gruesome last half of third tape.   ©1984

198

ADVENTURES OF CHARLIE WANDERMOUSE:
Tape 1

Content: ★★1/2
Technical Quality: ★★★

PRIMARY AUDIENCE: Ages 3 - 8
STYLE: Animation

LENGTH: 29 min.
DISTRIBUTOR: Brownlow Publishing Co.
817/831-3831

This tape contains two stories. Each retells a Bible parable but with a contemporary flare. Both stories are led into and summarized by a short discussion between three puppet characters, with the story itself presented in colorful animation. This first tape in the Charlie Wandermouse series introduces the star, Charlie Wandermouse (a mouse character that is described as a world-famous musician, traveler, and explorer) and his always present writer/friend, Professor Scribbler. The two stories on this tape are LEARNING ABOUT GOD'S FORGIVENESS (a take-off on the parable of the Prodigal Son of Luke 15 that encourages young viewers to say they are sorry when they do something that disappoints their parents) and LEARNING HOW TO BE A REAL FRIEND (similar to the Luke 10 story of the Good Samaritan; challenges viewers to be a friend to people who are in need). Quality of animation is superior to the short puppetry sequences. ©1989

AIDS: A CHRISTIAN PERSPECTIVE

Content: ★★★
Technical Quality: ★★

PRIMARY AUDIENCE: Jr. High - Adult
STYLE: Interview

LENGTH: 26 min.
DISTRIBUTOR: Bridgestone Group
619/431-9888

Even though medical knowledge and statistics relative to the AIDS virus are revised almost monthly, this video contains enough basic (non-time sensitive) material to assure that it will be useful through the early 1990s. Main content is Dr. C. Everett Koop (past Surgeon General of the USA) sitting at his desk talking about what causes AIDS and how to minimize one's chances of contracting it. The video's moderator, author/psychologist Dick Day, attempts to explain *why* people continue to involve themselves in homosexuality and I.V. drug use when they know they could contract AIDS. The tape endorses morals-based AIDS education and much greater involvement in the lives of AIDS patients by Christians. 1987

THE AMAZING BOOK

Content: ★★★★★
Technical Quality: ★★★★★

PRIMARY AUDIENCE: Ages 3 - 10
STYLE: Animation

LENGTH: 25 min.
DISTRIBUTOR: Multnomah Press
503/257-0526

When it comes to assembling a home video library, this one is an essential! Excellent quality animation *and* message (a combination not found often enough in children's videos). This is a spellbinder that should mesmerize every child over and over again. Instills a genuine interest in and hunger to know the Bible. Includes six sing-a-long songs that get even mom and dad tapping their feet. Deals with how we got the Bible, interesting stories in the Bible, memorizing the books of the Bible, Bible trivia, famous men who loved the Bible, and the timelessness and applicability of the Bible. Uses animated mice characters (Revver and Doc Dickory) and a mole (Dewey Decimole) to move into each subject and song.

©1988

AMY GRANT: "Find A Way"

Content: ★★
Technical Quality: ★★★★★

PRIMARY AUDIENCE: Jr. High - College
STYLE: Music video

LENGTH: 20 min.
DISTRIBUTOR: Word, Inc.
214/556-1900

To those who like Christian rock music and the rapid fire visual style of MTV, "Find A Way" probably rates five stars for content. However, to those who don't follow the subtle messages behind the harder to figure out scenes (or who question whether there really is any redeeming message at all), one or two stars are quite generous. Whatever *your* content rating (and it will undoubtedly vary widely depending on your age, pre-conversion lifestyle, present music taste, etc.) there is no question as to the level of technical expertise with which the five music videos on this tape were produced. Top notch, technically speaking! Personally, however, this reviewer couldn't find much meat to chew on in the message of the songs. (A good alternative to secular rock videos, but mom and dad may have to grin and bear it a little just the same.)

©1985

Animated Stories from the New Testament:
Vol. 1—THE KING IS BORN

Content: ★★1/2
Technical Quality: ★★★★★

PRIMARY AUDIENCE: Entire family
STYLE: Animation

LENGTH: 26 min.
DISTRIBUTOR: Family Entertainment
Network
214/341-8518

This is an animated re-telling of the birth of the baby Jesus. Unlike Hanna-Barbera's widely distributed "Greatest Adventure" series, no modern characters are included to interfere with the Bible story. However, the scriptwriters still have trouble properly representing some very straightforward scriptural details. Examples: Only one angel is shown at the announcement to the shepherds of Jesus' birth; no reference to the Matthew 1:19-25 account of Joseph's intention to "divorce Mary quietly" before an angel came to him and assured him of Mary's favor with the Lord. Animation, editing, sound effects, music, and all other technical details are top quality, but viewers should have their Bibles handy to recognize script *vs.* Scripture discrepancies, and to point them out to younger children.   ©1988

Animated Stories from the New Testament:
Vol. 2—HE IS RISEN

Content: ★★★
Technical Quality: ★★★★★

PRIMARY AUDIENCE: Entire family
STYLE: Animation

LENGTH: 29 min.
DISTRIBUTOR: Family Entertainment
Network
214/341-8518

This technically superb video sensitively portrays the emotions of the Resurrection story. Depicts the fear and remorse of the disciples after His death. Gives an account of the angel's appearance to the Roman guard at Jesus' tomb and then of Jesus' appearances to all of His disciples during the weeks prior to His ascension. Regretfully, the *writers* of this video were not as attentive to details as were the skilled animators and musicians. Some Easter story Scriptures contradicted here include John 20:5-7, 21:4,12; Matthew 28:11. The basic story of Christ's death and resurrection is communicated in a very entertaining manner, but viewers would do well to keep a Bible handy to point out the minor content inaccuracies.   ©1988

Animated Stories from the New Testament:
Vol. 3—THE PRODIGAL SON

Content: ★★★★
Technical Quality: ★★★★★

PRIMARY AUDIENCE: Entire family
STYLE: Animation

LENGTH: 27 min.
DISTRIBUTOR: Family Entertainment
Network
214/341-8518

The whole family will enjoy this highly but logically embellished version of the story of the Prodigal Son (from Luke 15:11-32). In this parable a young man asks for his share of his father's estate prior to his father's death, goes to the big city where he squanders everything, returns destitute, and is lovingly welcomed back by his very concerned father. The Disney-like animation is excellent—bright, colorful, and with good attention to detail. Excellent theme song. The story itself is well written, emotion charged, and includes several statements and actions which are not specifically recorded in Luke but which are in keeping with the tone of the actual scriptural account. Noticed only one statement contrary to the Luke account, but that was minor. (Upon return of the son and his father's joyous welcome, frustrated older brother jealously refers to the roasting of a *lamb* rather than a *goat*, as is recorded in Luke 15:29.) Worthy of repeated viewings. ©1988

Animated Stories from the New Testament:
Vol. 4—THE GOOD SAMARITAN

Content: ★★★1/2
Technical Quality: ★★★★★

PRIMARY AUDIENCE: Entire family
STYLE: Animation

LENGTH: 27 min.
DISTRIBUTOR: Family Entertainment
Network
214/341-8518

Based on the familiar, emotionally touching story of THE GOOD SAMARITAN found in Luke 10:25-37, this beautifully animated video is very well written and benefits from an excellent musical score. Includes lots of logical extra-biblical embellishments, but also some which hinder rather than aid in communicating this important story. (The two thief characters are funny but too unrealistic. They detract slightly from this otherwise believable account.) A few minor details of the Luke 10 account were altered (shows Samaritan riding a horse rather than a donkey, doesn't allude to robbed man's clothes being stolen, Bible says Samaritan gave innkeeper only two coins, etc.) but nothing of great spiritual or doctrinal significance. With Disney-like quality, this video succeeds well in communicating our responsibility to reach out in love to those in need. ©1989

Anthony Paul's - SHARING AND KINDNESS

Content: ★★★★★
Technical Quality: ★★1/2

PRIMARY AUDIENCE: Ages 2 - 5
STYLE: Filmation

LENGTH: 24 min.
DISTRIBUTOR: Multnomah Press
503/257-0526

This is one of several tapes in Anthony Paul's "Character Builder Storybook" series. Consists of two stories per video, each using message-based sing-along songs to develop and reinforce desirable character traits. The colorful but very simple animation (pans and zooms of non-moving cartoon characters) is best suited to very young audiences. Low price and excellent content make the series a great value. The two stories on this tape cover the traits of SHARING and KINDNESS. In "Benny Loves Fiddleberries," a young bear selfishly builds a dam to enable him to gorge on fiddleberries at the expense of other hungry bears downstream. The young bear soon recognizes his thoughtlessness and tears down his dam. The second story, "The Trouble With Tuffy," communicates the benefits of treating others as you would want them to treat you. The neighborhood bully (Tuffy) loses his beloved dog and little Stevie and Nancy show him kindness by helping him to earn enough money to get the dog out of the pound. ©1988

BASICS OF BASS FISHING

Content: ★★★★
Technical Quality: ★★★★

PRIMARY AUDIENCE: Fishermen, all ages
STYLE: How-To

LENGTH: 35 min.
DISTRIBUTOR: Ken Anderson Films &
Video
219/267-5774

This is a bass fisherman's "how-to" tape which includes a good spiritual application of how to deal with failure and frustration. Dr. John Davis, Christian college president and author of numerous books and articles on outdoor topics, demonstrates his skill and knowledge as a widely recognized expert in bass fishing. Actual fishing footage, plus a studio-sized miniature model of a lakebed help the viewer understand where to cast for "Ol' Bucketmouth." Good underwater footage of largemouth bass and the action of various lures. Great for the beginning and intermediate bass fisherman. It visually answers such questions as how to select the proper type of rod and reel, arrange your tacklebox, select line, cast with accuracy, and how to select from the wide variety of available artificial lures. Excellent gift for any fisherman or to help better prepare and inspire those whom you would like to accompany you on your next trip to the lake. *(Also available on 16mm film.)* 1987

BEAR HUGS: Volume 2

Content: ★★★
Technical Quality: ★★★

PRIMARY AUDIENCE: Ages 2 - 7
STYLE: Animation and real-life

LENGTH: 25 min.
DISTRIBUTOR: Brownlow Publishing Co.
817/831-3831

This animated children's video contains two separate stories. Each is taken from the BEAR HUGS book series. In the first segment, "Bearing Fruit," a short, real-life dramatic sequence between a father and his son and daughter discussing their fresh basket of fruits leads into an animated story that uses teddy bear type cartoon characters to introduce children to the concept of the Fruit of the Spirit (Galatians 5:22-23). Symbolism of the fruit in this first story is probably above the comprehension level of most children in the intended age group, however. The second segment, "Bear Buddies," uses the same style to develop a story based on Proverbs 17:17. Very understandable for intended audience. Reinforces parents' admonitions for children to share their belongings and to treat others kindly. Helps children understand how to make and keep friends. The quality of animation is very simple. Characters move, but only slightly. Most "movement" is dependent on dissolves between pictures. Good narrator and lively background music.   ©1987

BEYOND THE NEXT MOUNTAIN

Content: ★★★★★
Technical Quality: ★★★★1/2

PRIMARY AUDIENCE: Jr. High to Adult
STYLE: Drama

LENGTH: 1 hr., 37 min.
DISTRIBUTOR: Word, Inc.
214/556-1900

In 1905 a 22-year-old chemist ventured into the dangerous mountains and jungles of a northeastern India tribe of headhunters. He lived among them, learned their language, and shared the gospel message with them. A couple years later, he was unwillingly forced to return to Canada. However, during succeeding years, the seeds that he had sown made possible a tremendous harvest of souls. This challenging, heartwarming film follows the life and spiritual growth of Rochunga Pudaite, son of one of the Canadian missionary's few converts. It dramatically shows how Rochunga leaned on God to overcome tremendous hardships to educate himself so that he could translate the entire New Testament into the language of his people (a language which had never been written down before!). Rochunga's attitudes and actions are an inspiration to all who seek to follow Jesus Christ. Technical aspects, including the excellent photography, authentic costumes, sets, etc., provide a good feel for India's culture and customs. View this one with your family on video, or, better yet, convince your pastor or Sunday School teacher to rent the 16mm version to show to your entire church! *(Also available on 16mm film.)*   ©1981

BORN AGAIN

Content: ★★★★
Technical Quality: ★★★★★

PRIMARY AUDIENCE: Jr. High - Adult
STYLE: Drama

LENGTH: 1 hr., 50 min.
DISTRIBUTOR: Nelson Entertainment
213-285-6000

During Richard Nixon's last two years as President, Charles Colson was best known for his involvement in the Watergate break-in. In BORN AGAIN, Dean Jones plays the role of Colson and is joined by other top-quality actors to re-enact the circumstances that led to Colson's salvation, his subsequent jail term, and his experiences while in prison. Shows the corruptive influence of unbridled power and reveals how—while in prison—Colson desperately struggled with unanswered prayer. Upon his release, he recognized that God had, in fact, heard his prayers and that all things work together for good to those who love God. The realism and warmth of this story add a nice touch to the technical excellence with which it was produced. Too frequently peppered with profanity (though realistic, 25 instances is a bit much) to be widely used in Christian churches. Too bad, because the basic story is very good.   ©1978

CARMAN: Radically Saved

Content: ★★★
Technical Quality: ★★★★

PRIMARY AUDIENCE: Jr. High - College
STYLE: Live concert

LENGTH: 1 hr., 11 min.
DISTRIBUTOR: Benson Company
615/742-6800

This Carman concert was recorded live and later edited and embellished with high quality title sequences. While some of this popular Christian performer's songs and stories will come across as borderline sacrilegious and frothy to many parents, it's intended audience (teens and young adults) will enjoy it as clean, wholesome entertainment with an emphasis on having fun. The music is upbeat, the stories are funny (though not great examples of solid theology), and the gradual progression from an emphasis on humor and entertainment to worship and praise and finally to a call to revival is masterfully carried out. Ten songs, including: "Radically Saved," "Revive Us Again," and "The Champion." Simple staging with modest lighting effects. No on-stage band so viewers are not distracted by undue showiness.   ©1988

CAUGHT

Content: ★★★★1/2
Technical Quality: ★★★★★

PRIMARY AUDIENCE: High school - Adult
STYLE: Drama

LENGTH: 1 hr., 55 min.
DISTRIBUTOR: World Wide Pictures
612/338-3335

In this exciting, high energy drama 18-year-old Tim Devon accidentally learns that his birth was the result of a premarital college affair. Angry at his mother and desperate to learn more about his true father, Tim goes to The Netherlands to search. With only an old yearbook picture and a name, he soon runs out of leads and becomes trapped in a life of selling drugs, stealing, and male prostitution to "earn" money. Into Tim's hellish life walks the refreshing Rajam. Rajam is a naive, tenderhearted, itinerant evangelist from India. He repeatedly shows exemplary Christian love to the very unlovely Tim. The cold, worldly reality of Tim's circumstances contrast strongly with the warmth of Rajam's deeply committed life. Excellent! Scenes of harsh violence and clear allusions to male prostitution earned this show a PG-13 rating. Not recommended for pre-teens but a high quality motion picture with a good gospel presentation and challenge to Christians to pursue non-Christians for the glory of God. Impressed by the power of one person caring deeply about another, this is one story you may never forget. *(Also available on 16mm film.)*   ©1987

CHILDREN'S VIDEO BIBLE: The Beginning

Content: ★★★
Technical Quality: ★★1/2

PRIMARY AUDIENCE: Ages 4 - 7
STYLE: Filmation

LENGTH: 50 min.
DISTRIBUTOR: Bridgestone Group
619/431-9888

THE BEGINNING is the first of five watercolor illustration-based videos in the "Children's Video Bible" series. Contains Bible-based short stories that are each about 5 minutes in length and are taken directly from Lion Publishing's popular children's book, THE LION STORY BIBLE. While neither the book or video are designed to present a thorough account of each Bible story, most of the important highlights are included. The stories are presented in chronological order, providing children with a better understanding of the sequence and inter-relationship of various events of Old Testament history. Although the brevity of each story provides a good overview of Scripture, it is also a hindrance. Five minutes per story is not enough time to include or develop some very important specifics. Also, while the illustrated characters are good and generally life-like, they do not actually move. Rather, an illusion of movement is created by the camera's pans and zooms.   ©1988

CHRISTMAS IS

Content: ★★★
Technical Quality: ★★★★

PRIMARY AUDIENCE: Ages 3 - 10
STYLE: Animation

LENGTH: 18 min.
DISTRIBUTOR: Family Films
314/664-7000

In this fully animated children's drama a young boy, Benji, is frustrated because, for the second year in a row, he has been given the "unimportant" role of the second shepherd in his school's Christmas play. While reading through his script at home, Benji imagines that he and his dog Waldo are present at the birth of Jesus 2,000 years before. After speaking with the "real" second shepherd, Benji decides that his part in the play is important after all. Benji ends up playing his part with enthusiasm. Entertaining. No scriptural inaccuracies. A good teaching tool or discussion starter for family devotions or for children's Sunday School, youth group, etc., since it is cleared for these types of "public-use" settings. *(Also available on 16mm film.)*

©1974

CIRCLE SQUARE: Volume 12
"Broken Mirrors"

Content: ★★★
Technical Quality: ★★

PRIMARY AUDIENCE: Ages 4 up
STYLE: Educational

LENGTH: 45 min.
DISTRIBUTOR: Tyndale Christian Video
312/668-8300

Subtitled *"I Only Feel Handicapped When You Treat Me That Way,"* this is the twelfth in the Circle Square Gang home video series. The cassette actually has two programs on one tape. One is about respecting the handicapped, and one is about the importance of learning and being prepared to use CPR. The program about CPR in itself makes the tape well worth viewing and should be seen by the whole family several times. The main program (treating the handicapped normally) is an encouragement to strive to be the best that you can be. Shows a group of leg and arm amputees learning how to snow ski and to overcome their handicaps. ©1985

CITY OF THE BEES

Content: ★★★★★
Technical Quality: ★★★★

PRIMARY AUDIENCE: All ages
STYLE: Documentary

LENGTH: 28 min.
DISTRIBUTOR: Moody Institute of
Science
312/329-2190

This classic contains information that all age groups will enjoy. Using very appealing and educational close-up photography, CITY OF THE BEES reveals how bees guard their hives from intruders, how a queen bee is selected, how bees communicate by using sounds and dances, that drones (male bees) are killed and discarded immediately after the mating period is over, how many bee miles and bee hours it takes to build a honeycomb, etc. Features narration and short closing "sermon" by the late Dr. Irwin Moon. Dated by the early 1960s automobile models and a few other features, but not to the detriment of the excellent content in any way. Children will be fascinated by this production, and mom and dad will be very happy to allow them to see it again and again! *(Also available on 16mm film.)*   ©1962

THE CITY THAT FORGOT ABOUT CHRISTMAS

Content: ★★★
Technical Quality: ★★★★

PRIMARY AUDIENCE: Ages 4 - 12
STYLE: Animation

LENGTH: 28 min.
DISTRIBUTOR: Family Films
314/664-7000

THE CITY THAT FORGOT ABOUT CHRISTMAS teaches that "it isn't the decorations and celebrations that make Christmas important, but the love of God who sent the baby Jesus." A real-life modern grandfather shares with his grandson the story of a gloomy city that did not celebrate Christmas any longer. Animation then carries viewers back in time to the streets of that city. The people of the town didn't realize the joy that they were missing until Matthew the carpenter came to town. Matthew wins the town's friendship and finally shares with them that the birth of Jesus is the whole reason for the celebration of Christmas. Regretfully, the writers of this otherwise very good story do not even touch on the reason *why* Jesus' human birth was necessary. Parents should be sure to point out to children that the whole reason why Jesus became a human was to provide a way for individuals to be saved. Jesus' divine-humanity is nice, but pointless when this important fact is overlooked. *(Also available on 16mm film.)*   ©1974

COACH

Content: ★★★★
Technical Quality: ★★★1/2

PRIMARY AUDIENCE: Jr. High - Adult
STYLE: Drama

LENGTH: 1 hr., 18 min.
DISTRIBUTOR: Mark IV Pictures Video
515/278-4737

COACH is the story of a young science teacher (Stewart Linley) who reluctantly accepts the responsibility of coaching a very undisciplined basketball team at the Christian high school where he teaches. The "team" is the laughing stock of their conference, having won only two games in the last three years. Coach Linley's dedication to the task of teaching this ragtag group to strive toward excellence for God's glory pays off as the individual players learn to control their tempers and cooperate on the court. Coach Linley also emphasizes the applicability of God's Word to everyday life. While the movie gets off to a slow start and contains a couple of mediocre actors, it is very good overall and is built upon an excellent story sure to have a positive impact on its viewers. *(Also available on 16mm film.)* ©1982

THE CROSS AND THE SWITCHBLADE

Content: ★★★★
Technical Quality: ★★★★★

PRIMARY AUDIENCE: Jr. High - Adult
STYLE: Drama

LENGTH: 1 hr., 45 min.
DISTRIBUTOR: Vision Video
215/584-1893

This film was made several years before most of the young people who now view it were even born, but it exposes a drug culture and inner-city racial conflict similar in many respects to the situation of the present. Rated PG for language, violence, and drugs. THE CROSS AND THE SWITCHBLADE is based on a true story and stars Pat Boone (as David Wilkerson) and Erik Estrada (as Nicky Cruz). While some of the scenes of gang warfare go on longer than necessary, and while there are six instances of cursing, it is a realistic representation of the depravity and hardness of the hearts of man. It is also an encouraging and challenging enactment of how God can use willing servants in His plan to renew lives that the rest of society has written off as worthless and "beyond saving." Excellent in acting and in all technical aspects. A great film to view with neighbors or non-Christian friends in the proper age category. *(Also available on 16mm film.)* ©1970

A CRY FOR FREEDOM

Content: ★★★★★
Technical Quality: ★★★★1/2

PRIMARY AUDIENCE: Jr. High - Adult
STYLE: Documentary

LENGTH: 40 min.
DISTRIBUTOR: Gospel Films
616/773-3361

This award-winning documentary takes an honest and sometimes startling look at the drug problem facing youth from virtually all social and economic segments of society. Focuses on cocaine and crack. Moderated by a college-aged young man, A CRY FOR FREEDOM is fast-paced, challenging, and eye opening. Especially well suited for a teenage and young adult audience, the film was produced in association with David Wilkerson's tremendously successful drug rehabilitation ministry, TEEN CHALLENGE. Includes interviews with the "Globetrotters" Meadowlark Lemon, the "Beach Boys" Mike Love, a Hollywood stuntman, a model, and various medical doctors. Current and past drug users describe both the "positive" and negative sides of using cocaine and its hybrid, crack. Addictive and other detrimental effects are described in no uncertain terms. Closes with several short testimonies from past coke users. Should be available as a resource tool in every church's video lending library. *(Also available on 16mm film.)* ©1987

CRY FROM THE MOUNTAIN

Content: ★★★★1/2
Technical Quality: ★★★★★

PRIMARY AUDIENCE: Entire Family
STYLE: Drama

LENGTH: 1 hr., 16 min.
DISTRIBUTOR: World Wide Pictures
612/338-3335

In this action-packed drama a father (Larry) and son (Cal) fly deep into the Alaskan wilderness to share a special kayak trip. Early on the trip Cal learns that his father and pregnant mother (Carolyn) are planning a divorce, due chiefly to Larry's infidelity. The next day Larry is seriously injured while he and Cal attempt to navigate some rapids. With the help of a hot-tempered mountainman, Larry is rushed by rescue helicopter to a hospital. There he asks Carolyn to reconsider her decision to divorce. He promises his deepest love to her. (Carolyn has been strongly considering aborting their unborn child while Larry and Cal have been gone.) Through a well-scripted turn of events, Carolyn, Cal, and the mountainman all end up attending a Billy Graham crusade where they each receive Christ as Savior. A beautiful scene of Carolyn's acceptance of Christ and His love is portrayed as she returns to Larry's hospital room and tearfully tells her unconscious husband that she now has new hope for their marriage. Excellent in all aspects. Rich music and beautiful wilderness photography. Great viewing for the entire family! Buy this one for loan-outs through your church's video library. *(Also available on 16mm film.)* ©1985

C.S. LEWIS: "Through The Shadowlands"

Content: ★★★★
Technical Quality: ★★★★★

PRIMARY AUDIENCE: High School - Adult
STYLE: Docu-drama

LENGTH: 1 hr., 13 min.
DISTRIBUTOR: Vision Video
215/584-1893

C.S. Lewis's writings are some of the most popular and enjoyable works of Christian fiction ever published. But most Lewis fans probably are not aware of the near faith-shattering experience that he faced after most of his works had already been released. In this top quality drama, we see how Lewis meets Joy in a secret, name-only wedding that is never consummated. However, the two adults' relationship blossoms. When it is discovered that Joy has cancer, Lewis marries her again, this time in her hospital room "before God." Painfully grief stricken by Joy's eventual death, Lewis seriously questions his own basic Christian beliefs. He questions God's love and exhibits his anger at Him. Eventually, Lewis regains faith in his own oft used statement that this earthly life is but a "shadowlands" of what is to come. Especially meaningful to (and potentially hard on) anyone who has ever lost a loved one to an extended illness. *(Also available on 16mm film.)* ©1985

DANGEROUS JOURNEY

Content: ★★★★★
Technical Quality: ★★★★

PRIMARY AUDIENCE: Ages 8 - Adult
STYLE: Filmation

LENGTH: Nine 15-minute segments on 2 tapes
DISTRIBUTOR: Vision Video
215/584-1893

An animated version of John Bunyan's immortal classic, PILGRIM'S PROGRESS, this is an excellent series to use for family devotions. A very well-written story with tremendous spiritual symbolism, DANGEROUS JOURNEY introduces viewers to basic themes of the Christian life. Uses allegory, animation, and a compelling sound track to depict many of the challenges of a life committed to Jesus Christ. Animation is simple—pans and zooms of excellent illustrations. No movement of characters whatever. Narrator does almost all character voices—very good. This series should be seen and discussed by every Christian family! Sponsor this one for your church's video lending library. *(Also available on 16mm film.)* ©1986

DAVEY AND GOLIATH: Volume 1 (of 25)

Content: ★★★
Technical Quality: ★★★1/2

PRIMARY AUDIENCE: Ages 3 - 10
STYLE: Claymation

LENGTH: 29 min.
DISTRIBUTOR: The Program Source
313/333-2010

This tape contains two of the many DAVEY AND GOLIATH episodes that were originally broadcast on television during the 1960s and early 1970s. Employing a unique form of clay animation, the creators of this series were successful in developing characters (young Davey Hanson, his talking dog Goliath, and Davey's family and friends) whose entertainment appeal and on-screen actions have survived almost three decades of cultural and technological change. In the first of two well-written stories on this tape, "Lost In a Cave," Davey learns that it is important to obey the instructions of adults and that God really does care about what happens to Him. In the second story, "Cousin Barney," Davey, and his sister Sally learn from their father that we should love everyone, even if they're a little clumsy (like young cousin Barney) because "God loves all of us!" mid-1960s

---

DAVEY AND GOLIATH: Halloween Who-Dun-It

Content: ★★★1/2
Technical Quality: ★★★★

PRIMARY AUDIENCE: Ages 4 - 10
STYLE: Claymation

LENGTH: 29 min.
DISTRIBUTOR: Gospel Films
616/773-3361

The DAVEY AND GOLIATH series can be counted upon to include Christian values without sounding preachy. This episode is no exception. Davey, sister Sally, and Goliath the talking dog all dress up for a Halloween costume contest and then for trick-or-treating. Davey becomes very secure in thinking that he can be mischievous without being found out because no adults are able to tell who he is when dressed up in his homemade "Man from Mars" outfit. He even ends up destroying a beehive owned by one of his adult friends. The next morning Davey feels remorseful and realizes that God knows who destroyed the hive even if the owner does not. He ends up confessing and apologizing to the owner. The story is dated and somewhat misleading by the fact that young children are freely allowed to trick-or-treat after 10:00 p.m. without an adult chaperone. ©1967

DAVEY AND GOLIATH: Happy Easter

Content: ★★★
Technical Quality: ★★★★

PRIMARY AUDIENCE: Ages 4 - 10
STYLE: Claymation

LENGTH: 30 min.
DISTRIBUTOR: Gospel Films
616/773-3361

This "special" episode in the series deals with the death of a loved one and the promise of resurrection for those who are in Christ. Davey has a special affection for his grandmother, a unique woman who lets him help her bake cakes, practices baseball with him, and doesn't get angry when he accidentally breaks an attic window pane. ("Don't worry Davey, we'll fix it together.") To everyone's surprise, the grandmother dies unexpectedly. Davey is heartbroken, but later becomes happy for his grandma when he is reminded through viewing the rehearsal of an Easter pageant that he will be with her again one day, thanks to Jesus' death and resurrection. Pastor's comforting words at graveside service "where He is all of us will be" assumes that viewers are already Christians, since there is no mention anywhere in the film of need for repentance from sin and acceptance of Jesus Christ as Savior to receive the free gift of salvation. mid-1960s

DESTRUCTIVE DAVID

Content: ★★★★
Technical Quality: ★★★★

PRIMARY AUDIENCE: Ages 3 - 10
STYLE: Animation

LENGTH: 25 min.
DISTRIBUTOR: Family Films
314/664-7000

If you are in the process of trying to guide an active young boy through the "seek and destroy" stage of childhood, you will certainly appreciate this animated story. David squashes daisies, maliciously stomps through ant hills, and destroys beautifully woven spider webs. It is through a visit with a talking Peacock statue at an outdoor art fair that David learns why he should care about such "useless" things. The message is good: we should appreciate and care for God's creative beauty and recognize His love in sending His Son. Music and songs complement the story well. Worth replaying often even if just to sing-along with the songs. Don't pigeonhole this tape as one for terrorizing little boys. It is also great to use as preventative medicine in those families who enjoy the presence of "little angels," too. ©1988

213

A DISTANT THUNDER

Content: ★★★★
Technical Quality: ★★★1/2

PRIMARY AUDIENCE: Jr. High - Adult
STYLE: Drama

LENGTH: 1 hr., 17 min.
DISTRIBUTOR: Mark IV Pictures
515/278-4737

This drama, the second in a four-part series, begins where its predecessor (A THIEF IN THE NIGHT) ended. An extremely challenging production, both emotionally and spiritually. Acting and technical aspects are generally good, only infrequently distracting from the well-written story of Patty, a young woman living in the "end times" of the Great Tribulation referred to in biblical prophecy. She and two friends live as fugitives, constantly pursued and finally captured by the evil forces of UNITE. Inevitably, they must choose between taking "the mark of the Beast" or facing a terrifying death. This entire series brings up strongly debated theological interpretations regarding a pre-tribulation rapture, the Tribulation, and end times. While the viewer may not agree with or be clear on all points raised, the film is extremely stimulating—challenging both Christians and non-Christians to explore God's Word for a deeper understanding of Scriptures. Also challenges the believer to share the gospel message with others. Not recommended for pre-Jr. High ages due to level of intensity. *(Also available on 16mm film.)*   ©1977

DISTINCTIVELY HUMAN

Content: ★★★★★
Technical Quality: ★★★★★

PRIMARY AUDIENCE: Ages 10 - Adult
STYLE: Documentary

LENGTH: 58 min.
DISTRIBUTOR: Moody Institute of
Science
312/329-2190

From the attention-grabbing introduction straight through to its mild evangelistic conclusion, this documentary is a joy to watch! By combining mind-boggling statistics and information with excellent visual footage, DISTINCTIVELY HUMAN reveals how every cell, every nerve, every bone in the human body points unmistakably toward the Master Craftsman who created us all. Subjects covered include footage showing the growth of a baby beginning only minutes after conception, moving x-rays of various men and women performing different physical activities, film of the inside of an actual, functioning human heart, unique animation depicting the complexity of the human brain, etc. This award-winning production deserves a place in every home, church, and school video library! *(Also available on 16mm film.)*   ©1987

DOVETALE'S: Noah's Ark

Content: ★★★
Technical Quality: ★★

PRIMARY AUDIENCE: Ages 2 - 4
STYLE: Filmation

LENGTH: 40 min.
DISTRIBUTOR: Performance Unlimited
512/222-1877

This is the first of a 4-part series. Features simple animation from the award-winning DOVETALES book series. While the video is entitled NOAH'S ARK, it is actually loaded with twenty-one additional simplified Bible stories including the Creation, fall, Tower of Babel, Jacob and Esau, Joseph, Balaam's Donkey, etc. This video, both in visual quality (pans and zooms of cartoon characterizations) and in its superficial presentations of the Bible stories, is best suited for very young children up to about four years. For older children look for videos that reveal more of the details about each story such as Ken Anderson's "Draw-on" animation series, also reviewed in this book.

©1988

ECLIPSE OF REASON

Content: ★★★★★
Technical Quality: ★★★★

PRIMARY AUDIENCE: Jr. High - Adult
STYLE: Documentary

LENGTH: 27 min.
DISTRIBUTOR: Bernadell, Inc.
212/463-7000

This award-winning pro-life documentary opens with a short introduction by Charlton Heston who appeared without financial renumeration. Mr. Heston reprimands the television community for failing to better inform the public with regard to the abortion issue. The film is moderated by Dr. Bernard Nathanson, past director of the largest abortion clinic in the world—a man ultimately responsible for over 75,000 abortions. ECLIPSE OF REASON focuses on "late abortions," those performed after the first three months of pregnancy. Reveals many startling statistics, and graphically documents the actual abortion of a five-month-old boy, pulled in pieces through the uterus. Also includes interviews with women who have been physically and emotionally scarred by their own abortions. Due to the graphic nature of the material presented, the best uses of this very powerful film would be to men, and to women who have not had an abortion previously. A sobering but excellent resource for the church's video library. *(Also available on 16mm film.)*

©1987

215

THE EVOLUTION CONSPIRACY

Content: ★★★★
Technical Quality: ★★★★

PRIMARY AUDIENCE: Jr. High - Adult
STYLE: Documentary

LENGTH: 56 min.
DISTRIBUTOR: Bridgestone Group
619/431-9888

While it is subtitled "A Quantum Leap Into The New Age," THE EVOLUTION CONSPIRACY really focuses on presenting a concise overview of the modern Creation/evolution controversy. This documentary points out that no truly valid transitional forms ("missing links") between various types of animals and man have ever been found in the earth's fossil record. Also briefly discusses the theory of "punctuated equilibrium," the Scopes Monkey Trial, and other related topics. Very summarily shows the connection between Darwinian evolution and the New Age movement. "Evolution, whether biological or mystical, is man's way to explain away God and His creation, and put man in God's place." Includes short interviews with many different creationists, evolutionists, and New Agers. Pacing is very good. Rapid transition from one visual to the next. While it does not develop the many evidences for creation, this film does succeed in packing a wide variety of information into one hour. For more depth and specific evidences in support of scientific creation, see the six-part ORIGINS—HOW THE WORLD CAME TO BE series also reviewed in this book.

©1988

FIRST FRUITS

Content: ★★★★★
Technical Quality: ★★★★

PRIMARY AUDIENCE: Ages 10 - Adult
STYLE: Drama

LENGTH: 1 hr., 10 min.
DISTRIBUTOR: Vision Video
215/584-1893

This entertaining docu-drama is much more than just a historical look at the Moravian mission movement. It is a well-written and heartwarming challenge to deeper commitment in the Christian life and in our emotional, prayerful, and financial support of contemporary missionaries. With excellent cinematography, this multi award-winning production transports its viewers back in time more than 250 years. Viewers follow two young Moravian men who left their comfortable European community convinced that they were called of God to preach to the ill-fated/mistreated slaves of the West Indies. They were willing even to become slaves themselves. You will rejoice over the first fruits of their ministry and the worldwide impact of their commitment on the course of Christianity and world history. Merits repeated viewing. Should be part of every church and Christian school video library. *(Also available on 16mm film.)*

©1982

THE FLYING HOUSE: Volume 4

Content: ★
Technical Quality: ★★★1/2

PRIMARY AUDIENCE: Ages 5 - 8
STYLE: Animation

LENGTH: 39 min.
DISTRIBUTOR: Tyndale Christian Video
312/668-8300

THE FLYING HOUSE series is another in the genre where modern youths travel backward in time. Like cookies from the same mold, the main difference between this and other similar series (such as SUPERBOOK and GREATEST ADVENTURES) is simply the color of the frosting. All are cartoons based on popular Bible stories. All are intended for young viewers, and yet almost none is appropriate for repeated viewing due primarily to two reasons: 1) Children from 5 to 7 do not readily separate modern youths from actual Bible story, and, 2) the story is often diluted by liberal doses of "artistic license." In this volume, three children, a robot, a house, and a bumbling professor end up in two different time "zones." The first of two episodes on this tape, THE PRIZE THAT WAS WON AND LOST, deals with the beheading of John the Baptist. Biggest problem here is that the dancer who asked for John's head on a platter is portrayed as an innocent 10-year-old who is an unwilling participant in the matter. The second episode, ANOTHER LIFE, deals with Jesus' raising Jairus' daughter from the dead in Mark 5:22, Matthew 9:23, and Luke 8:41. Very entertaining but contains many inaccuracies. Don't show this series without also observing and discussing the differences from the biblical record with your children.   ©1982

GERBERT: "Safe in His Arms"

Content: ★★★★
Technical Quality: ★★★★

PRIMARY AUDIENCE: Ages 3 - 8
STYLE: Puppetry + Live action

LENGTH: 31 min.
DISTRIBUTOR: Word, Inc.
214/556-1900

In this fifth episode of the fast-paced GERBERT video series, the puppet star of the show, Gerbert, learns the value of protecting what is important. With a wide-ranging and quickly changing array of visuals (puppets, human actors, real-life outdoor photography, cartoon illustrations), the message of this tape also includes a special song and choreography sequence about how good policemen are to us, footage of skillful skateboarders demonstrating why protective gear is important, and a story that helps youngsters to appreciate unique people because God has a special plan for them, too. A nice addition to any church's video lending library. Bright colors, upbeat music, and fast editing are sure to captivate young viewers.   ©1988

217

GOD OF CREATION

Content: ★★★★
Technical Quality: ★★★1/2

PRIMARY AUDIENCE: Ages 8 - Adult
STYLE: Documentary

LENGTH: 28 min.
DISTRIBUTOR: Moody Institute of
Science
312/329-2190

This "Sermons From Science" video is hosted by the late Irwin Moon. Visually supported with high quality photography, Dr. Moon opens with an explanation of how many stars there are in the universe and then uses stories about lilies, caterpillars, and paramecium to illustrate that the same God who created the ten octillion (10,000,000,000,000,000,000,000,000,000) stars also created and cares deeply for each one of us individually. The footage showing a caterpillar's actual transformation into a chrysalis and finally into a butterfly is amazing! Although the tape is dated by the 1940s style of background music, GOD OF CREATION continues to minister to young and old alike. Challenging evangelistic ending by Dr. Moon concludes this unique and educational production. *(Also available on 16mm film.)* ©1945

GOD'S OUTLAW: The Story of William Tyndale

Content: ★★★★1/2
Technical Quality: ★★★★

PRIMARY AUDIENCE: High School - Adult
STYLE: Docu-drama

LENGTH: 1 hr., 33 min.
DISTRIBUTOR: Vision Video
215/584-1893

This intense drama is a true story, describing how a single individual can, by God's empowering, change the world for good. Well written and well acted, GOD'S OUTLAW also addresses international politics, false injustice, and a corrupt religious establishment. William Tyndale is burned at the stake at the end of the video. His crime—translating the Bible into English and publishing it for his fellow countrymen. High production values. A "must see" film for all Christian adults. Should be placed in every church's home video lending library! *(Also available on 16mm film.)*

©1987

GOLF FOR KIDS OF ALL AGES

Content: ★★★★
Technical Quality: ★★1/2

PRIMARY AUDIENCE: Ages 8 - 12
STYLE: How-to/Instructional

LENGTH: 53 min.
DISTRIBUTOR: Asian Home Video
404/442-1500

This is an excellent instructional tape for the junior or beginning golf enthusiast. Loaded with tips, it features PGA Tour veteran Wally Armstrong and his son, Scott. Also includes the animated "Gabby Gator." Wally creatively uses everyday objects and illustrations from other sports to aid in the effectiveness of this instructional video. Because Scott is left-handed, this production also demonstrates the feel and technique of a proper golf swing for left- and right-handed players. Helpful for both children and adults. Short salvation message at end of tape. A great gift for any sports-minded child.   ©1986

GOSPEL BILL: There Is No Such Thing as Monsters

Content: ★★★★
Technical Quality: ★★★

PRIMARY AUDIENCE: Ages 6 - 12
STYLE: Live action

LENGTH: 27 min.
DISTRIBUTOR: Infinity Video
918/585-5033

The GOSPEL BILL Show is a weekly children's program. Often peppered with elements of charismatic theology or worship style, it is carried on Saturday mornings on many Christian TV stations across the USA. "There Is No Such Thing as Monsters" is one of the weekly episodes and demonstrates biblical truths in a comical way by using an "Old West" studio setting and several humorous characters. In this episode, many residents of the town of Dry Gulch are afraid after hearing various monster stories. Mr. T.U. Tutwater initially laughs at the notion but soon is afraid even to go to sleep at night because he fears a monster is stalking him. Gospel Bill, the town sheriff, tracks the "monster" and discovers that it is really nothing more than a prankster dressed in a costume trying to scare his friends. Gospel Bill then tells the young viewers that they should be careful not to fill their minds with bad thoughts, but rather fill them with good thoughts so there is no room for any bad ones.

©1986

THE GREAT DINOSAUR MYSTERY

Content: ★★★★★
Technical Quality: ★★★

PRIMARY AUDIENCE: Ages 8 - Adult
STYLE: Documentary

LENGTH: 20 min.
DISTRIBUTOR: Films For Christ
602/894-1300

This unique "children's documentary" appeals to adults and children alike. Near timeless style effectively combines illustrations, simple animation, still photos and outdoor cinematography to reveal evidence that dinosaurs have lived at the same time as man "in spite of the fact that evolutionists dogmatically say it is impossible." One of the highlights is the actual photograph of a recently dead Plesiosaur carcass netted by a Japanese fishing vessel near New Zealand in 1977. This production shows how dinosaurs fit into and are mentioned in the Bible. Also uses accounts of secular writers and historians to document intriguing dinosaur sightings of only several hundred years ago. While it is used primarily by churches, THE GREAT DINOSAUR MYSTERY was produced with public school students in mind and is used in hundreds of public school science classes every year! *(Also available on 16mm film.)*
©1979

Hanna-Barbera's - THE CREATION

Content: ★★
Technical Quality: ★★★★★

PRIMARY AUDIENCE: Ages 5 and up
STYLE: Animation

LENGTH: 27 min.
DISTRIBUTOR: Sparrow Home Video
818/709-6900

Oh that more Christian producers had enough money to make their productions sparkle as well as the creators of the Hanna-Barbera series! However, the technical qualities of this tape, even the top-rate animation, do not make up for the many scriptural misrepresentations that young viewers will mistakenly understand as truth (especially if viewed repeatedly). Studies have shown that children from 5 to 7 do not readily differentiate reality from fiction. This is a well-produced entertainment tape, but it is not an accurate representation of scriptural narrative. Some of the problem areas include modern youths interacting with Adam and Eve, how Eve was presented to Adam, Adam "unknowingly" eating the forbidden fruit, rain in the Garden, no Cherubim, and no movement of God's flaming sword (stuck in the ground instead). Better to present a parable than so much inaccurate information to such young, naive eyes.
©1988

Hanna-Barbera's - DANIEL AND THE LION'S DEN

Content: ★1/2
Technical Quality: ★★★★★

PRIMARY AUDIENCE: Ages 5 and up
STYLE: Animation

LENGTH: 24 min.
DISTRIBUTOR: Sparrow Home Video
818/709-6900

This film shows Daniel being thrown into a den of hungry lions. Also includes Daniel's interpreting of the handwriting on the palace wall. The high quality animation is very entertaining and gives a good feeling of what the city of Babylon and its culture were probably like. However, the excellence ends with the animation. Too much attention is paid to the three fictional children that have passed through a "time-portal" to become a part of this story. They are shown directly interacting with Daniel, King Darius, and others and are repeatedly chased by soldiers, etc. These and other embellishments tend to detract from rather than enhance these important events of Bible history. Several minor points are overstated while important facets of the story have been excluded altogether. Viewing this video confuses even adults. Most of us simply are not well versed enough to be able to pick out what is misrepresented. Finally, many parents will not appreciate the scantily clad belly dancers who perform during a palace party.   ©1984

Hanna-Barbera's - DAVID AND GOLIATH

Content: ★★★
Technical Quality: ★★★★★

PRIMARY AUDIENCE: Ages 5 and up
STYLE: Animation

LENGTH: 25 min.
DISTRIBUTOR: Sparrow Home Video
818/709-6900

Three modern youths witness David and Goliath's confrontation after passing through a "time-portal." Excellent animation. However, like all other episodes in this series, this is a highly embellished representation of what is actually recorded in Scripture. In this story, the ground actually shakes when Goliath walks; Goliath is represented as three times taller than other men; modern youths interact with David, etc. While the writers included a reference by David that his reason for going to battle is "to uphold the name and honor of the Lord," the overall emphasis still focuses on David's skill rather than God's choosing and using of him (I Samuel 17:36-37). Oversights and unreasonable additions are generally negligible but viewers should reacquaint themselves with the biblical account prior to viewing since there are some minor discrepancies. Challenge your children to locate these, and then to describe them to you.   ©1984

Hanna-Barbera's - JOSHUA AND THE BATTLE OF JERICHO

Content: ★1/2
Technical Quality: ★★★★★

PRIMARY AUDIENCE: Ages 5 and up
STYLE: Animation

LENGTH: 25 min.
DISTRIBUTOR: Sparrow Home Video
818/709-6900

The basis for this animated adventure is found in Joshua 5:13—6:27. If the viewers are mature enough to be able to separate scriptural fact from scriptwriter's fantasy, they will find that the basic message is fairly close to the Old Testament account of this amazing story. However, the inclusion of three time-traveling youths seriously deters from the usefulness of the tape. One of the three youths gets captured and taken inside Jericho but is later rescued by his two cohorts after they sneak into the city by climbing over the walls at night. All three then escape from Jericho by "surfing" on a board through an underground river passageway. Exciting, but very misleading for younger viewers. Some of the biblical problems: 1) Jordan River simply dries up rather than being cut off and standing in a heap (see Joshua 3:13); 2) Rahab is portrayed as a wholesome young woman rather than as a harlot as stated in Scripture. Poor attention to important details. Not recommended.
©1985

Hanna-Barbera's - NOAH'S ARK

Content: ★★
Technical Quality: ★★★★★

PRIMARY AUDIENCE: Ages 5 and up
STYLE: Animation

LENGTH: 26 min.
DISTRIBUTOR: Sparrow Home Video
818/709-6900

This story of Noah and the ark is similar to the account recorded in Genesis, but certainly not with much attention to detail. The story is very entertaining, but is not appropriate for younger viewers since repeated viewing will ingrain inaccurate Bible knowledge. Only eight people survived this awesome, earth-changing event, not eleven! No appreciation for the years needed to build the ark is portrayed and certainly no feeling for the full year that man and animal shared the vessel. Unlike what is portrayed in this video, elephants and oxen most likely did not slide up and down the aisles of the ark, and the horn of a live hippopotamus was never used to plug a leak. Captivating storyline but poor and often completely unwarranted embellishments of history.
©1985

Hanna-Barbera's - SAMSON AND DELILAH

Content: ★★1/2
Technical Quality: ★★★★★

PRIMARY AUDIENCE: Ages 7 and up
STYLE: Animation

LENGTH: 27 min.
DISTRIBUTOR: Sparrow Home Video
818/709-6900

SAMSON AND DELILAH is based on chapters 13—16 of Judges. Shows the overwhelming physical strength and emotional frailty of this man whom God installed as Israel's leader for 20 years. While this episode does not distort the biblical record nearly as much as some of the other segments, it does go overboard in its extra-biblical embellishments. (One of the three modern youths who have passed through a "time portal" to Samson's period in history even builds and flies a hang-glider!) Also, the physical attributes of Delilah and other women are revealed more than many parents will appreciate. If you do rent or purchase this tape, you might play a family game by first reading aloud the entire Judges 13—16 account. Then, give each of your reading-aged children a Bible and challenge them to find the content problems in the video. In this way, the kids will become much more familiar with the true biblical account, and you will enjoy the family interaction and discussion.   ©1985

HEAVEN'S HEROES

Content: ★★★★
Technical Quality: ★★★1/2

PRIMARY AUDIENCE: Ages 8 - Adult
STYLE: Drama

LENGTH: 1 hr., 12 min.
DISTRIBUTOR: Mark IV Pictures Video
515/278-4737

In this true story, a series of flashbacks combine to reveal the life of Christian police officer Dennis Hill. Killed in the line of duty by a sniper, Officer Hill's marital and Christian life is an example and inspiration to all viewers. HEAVEN'S HEROES shows the problems and humanity of those who patrol our city streets and demonstrates effective lifestyle evangelism as Officer Hill strives to win his partner to Christ. Though the story is sad in that a loving father and husband is torn from his family and friends, the need for Christians to rely on Christ through rough times is well developed. A good drama to add to any church's video lending library. (*Also available on 16mm film.*)
©1980

THE HIDING PLACE

Content: ★★★★★
Technical Quality: ★★★★★

PRIMARY AUDIENCE: Jr. High - Adult
STYLE: Drama

LENGTH: 2 hr., 25 min.
DISTRIBUTOR: World Wide Pictures
612/338-3335

This intense story is a superbly produced drama about the lives, sufferings, and triumphant joys of the ten Boom family. Imprisoned in Nazi Germany's Ravensbrook concentration camp for helping to save the lives of Jews by harboring them in their Holland home, Corrie ten Boom and her sister, Betsie, suffer inhuman treatment at the hands of Nazi prison guards. The love between the two sisters and toward their God deepens as their trials intensify. Although THE HIDING PLACE contains some brief scenes of violence and some almost indiscernible cursing, it is extremely valuable and highly recommended. There is something very special and sobering about watching a show like this, always knowing that both the terrible sufferings and the triumphant joys were real. Try to provide an atmosphere as free from potential interruptions as possible. Doing so will greatly enhance this inspiring and timeless story. *(Also available on 16mm film.)*   ©1975

HOOMANIA

Content: ★★★★
Technical Quality: ★★★★

PRIMARY AUDIENCE: Ages 5 - 12
STYLE: Claymation/Drama combination

LENGTH: 37 min.
DISTRIBUTOR: Gospel Films
616/773-3361

This unique children's adventure combines real-life cinematography and colorful "claymation" to take a creative look at some simple lessons from the book of Proverbs. After accidentally breaking a living room window while playing baseball in his front yard, a young boy (Kris) escapes to the workshop of his elderly friend, the inventor. The elderly man offers Kris a chance to play an amazing board game that he has invented. The claymation portion of the video begins here as Kris accepts the offer and is immediately zapped into the game itself. As he strives to win the game by reaching Mt. Wisdom, he makes some very foolish choices, including keeping bad company and disregarding instruction. Kris is chased by a daffy Dodo Bird and an army of chessmen, and is tempted along his route by goodie gobbling sluggards. He finally learns from a wise old owl that wisdom means striving to please God. Awarded "Best Children's Film" by the Academy of Christian Cinemagraphic Arts. *(Also available on 16mm film.)*   ©1985

HOW SHOULD WE THEN LIVE

Content: ★★★★★
Technical Quality: ★★★★

PRIMARY AUDIENCE: High School - Adult

STYLE: Documentary

LENGTH: 30 minutes per episode
(10 episode series)
DISTRIBUTOR: Gospel Films
616/773-3361

HOW SHOULD WE THEN LIVE features the late Dr. Francis Schaeffer. It is a documentary series that contains a great deal of information about the philosophies that led to the degeneration of society today and challenges Christians with suggestions about how we are to respond to our contemporary world. Schaeffer moderates all ten episodes on-location in many different settings, principally throughout Europe. Includes a wide variety of illustrations and dramatizations and a unique style of photography. This is a series which requires serious viewers. Highly recommended and well suited for use in an adult Sunday School or Bible study classes, or for home viewing for personal enrichment. Nearly timeless in content. This series should definitely be a staple in every church and upper level Christian school video library. *(Also available on 16mm film.)*
©1977

HOW TO GET BETTER GRADES AND HAVE MORE FUN

Content: ★★★★1/2
Technical Quality: ★★★

PRIMARY AUDIENCE: Late High School - Adult
STYLE: Lecture

LENGTH: 59 min.
DISTRIBUTOR: Inspirational Media
714/886-5224

This motivational/how-to tape comes with a 100% money-back guarantee. If the owner's GPA has not improved after applying its study principles for one year, the full purchase price is refunded. Featuring M.I.T. graduate Steve Douglas, HOW TO GET BETTER GRADES teaches practical and proven study techniques that help students to improve their grades. It also shows business people how to develop some of the important skills necessary for a successful career. The production was shot in a TV studio with a live college-aged audience. The many colorful stories integrated by the speaker into his excellent teaching help to make this product worthy of gift-giving, especially to high school graduates who plan to continue their education. Includes a low-key evangelistic message.
©1986

HOW TO KNOW YOU'RE IN LOVE

Content: ★★★★
Technical Quality: ★★

PRIMARY AUDIENCE: Jr. High - College
STYLE: Lecture

LENGTH: 65 min.
DISTRIBUTOR: Word, Inc.
214/556-1900

This is part four of the lecture series *Straight Talk About Love, Sex, and Dating.* Shot before a live audience, popular youth speaker Dawson McAllister skillfully uses humor to reveal the difference between love and infatuation. He emphasizes that "love is both an emotional need and an act of the will based on a clear understanding of what the other person is like." Stresses to young men, "Don't get married unless you are willing to make your wife your second most important relationship" (God being #1). The message of this video is very good and worthy of being seen by every high school and college young person. The technical quality is satisfactory but very uncreative (really nothing more than a large group lecture sandwiched between a musical introduction and conclusion).

©1985

HUMAN RACE CLUB - Volume 1: "Self Esteem"

Content: ★★1/2
Technical Quality: ★★★★

PRIMARY AUDIENCE: Ages 6 - 10
STYLE: Animation

LENGTH: 22 min.
DISTRIBUTOR: Bridgestone Group
619/431-9888

This animated story, based on the book by Joy Berry, is the first in the HUMAN RACE CLUB series. "The Letter on Light Blue Stationery" deals with the subject of self esteem. Shows the dilemma of Pamela, a little girl who finds it hard to write a sympathy letter to the family of a classmate who has been killed suddenly in an automobile accident. Pamela is encouraged by her mother to mention several of her friend's outstanding qualities in the sympathy letter, but finds it hard to think of anything truly "special" about the average looking, slightly pudgy girl. It isn't until a rather somber talk with her friends in the neighborhood clubhouse that Pamela finally recognizes and truly appreciates the caring/helpful attitude of her friend and recognizes the value of every person. This story is very good. However, it lacks any reference to God or Scripture.

©1989

HUMPTY

Content: ★★★
Technical Quality: ★★★★1/2

PRIMARY AUDIENCE: Ages 3 and up
STYLE: Animation

LENGTH: 24 min.
DISTRIBUTOR: Bridgestone Group
619/431-9888

"HUMPTY is the animated story of an inquisitive egg who lived in a faraway land. Humpty discovers the value of good rules, but he learns it the hard way." Deceived by a dinosaur-like serpent into believing that he is the "best" egg, Humpty becomes conceited and mean to his neighbors. He eventually disobeys the king's rule against climbing over a tall, protective wall that surrounds his city. While Humpty learns that pride comes before a fall and that obedience is the real key to peace and contentment, the message of this film falls short in that the king (symbolic of God) freely forgives Humpty without any mention of atonement. *(Also available on 16mm film.)* ©1980

IMAGE OF THE BEAST

Content: ★★★★
Technical Quality: ★★★1/2

PRIMARY AUDIENCE: Jr. High - Adult
STYLE: Drama

LENGTH: 1 hr., 33 min.
DISTRIBUTOR: Mark IV Pictures
515/278-4737

The third of four films in Mark IV's dramatic "prophecy series," IMAGE OF THE BEAST attempts to depict the events and emotions of the mid-years of the seven year Great Tribulation. The anti-Christ has set up his one-world governmental system and anyone who has not taken his "mark" on the hand or forehead is hunted down and put to death. The first several minutes of this production are *extremely* intense—and not for young children or weak-stomached adults. Acting is good, but could have been assisted by stronger performances from several of the supporting actors. While the story and technical aspects may seem like science fiction at times, this is primarily because we cannot fathom the terrible reality of the tribulation to come. It is difficult to represent visually events that even scholars do not completely understand. This series is based upon a pre-tribulational rapture perspective. More than just good entertainment, this is a tremendous evangelistic tool. If possible, be sure to view A THIEF IN THE NIGHT and A DISTANT THUNDER before seeing this film. *(Also available on 16mm film.)*
©1981

ISRAEL - GOD'S CHOSEN LAND

Content: ★★★
Technical Quality: ★★1/2

PRIMARY AUDIENCE: Adults
STYLE: Documentary

LENGTH: 35 min.
DISTRIBUTOR: Christian Duplications
Int'l.
407/299-7363

By combining scenes of modern-day Israel, Scripture reading, cutaways to maps of Israel, historical facts told by an unseen narrator, and a good selection of background music, this rather low-budget production is a good tape for anyone who is considering a trip to, or who has already visited, "God's Chosen Land." A concise travelog of the attractions most frequented by North American visitors to Israel (including the traditionally accepted tomb of Jesus, the Garden of Gethsemane, the Jordan River, the temple site, etc.). Inclusion of aerial photography provides a good feeling for the geography of the area. Some scenes do not mesh with the narration or script reading, but good overall.

©1988

JESUS

Content: ★★★★★
Technical Quality: ★★★★★

PRIMARY AUDIENCE: All ages
STYLE: Drama

LENGTH: 2 hours
DISTRIBUTOR: Inspirational Media
714/886-5224

This film was made entirely in Israel with a cast of over 5,000 Israelis and Arabs. With a script taken entirely from the book of Luke, this is clearly one of the most accurate Bible dramas to date. The story begins with the angel's announcement to the virgin Mary of the impending birth of Jesus, visually depicts the main events of Jesus' life and ministry, portrays a painfully realistic crucifixion, and then ends with His ascension through the clouds. Includes none of the standard Hollywood sensationalism usually written into Bible films of this financial magnitude ($6 million). Wonderful attention to scriptural purity combined with excellent technical and cinemagraphic quality. Moves slower than North American viewers may be used to, but allows the many events depicted and message imparted to "sink in" rather than simply being lost as non-applicable entertainment. *NOTE: This film has been translated into almost 200 languages and is used as an evangelism tool throughout the world. Indoor and outdoor screenings of the 16mm version serve as the catalyst to introduce literally thousands each month to a personal relationship with Jesus Christ. (Also available on 16mm film.)*

©1979

JESUS' BICYCLE

Content: ★★★★★
Technical Quality: ★★★

PRIMARY AUDIENCE: Ages 4 - 10
STYLE: Drama

LENGTH: 21 min.
DISTRIBUTOR: Cathedral Films
818/991-3290

This children's drama is filled with the warmth of people helping other people. A mentally and physically handicapped teenage boy (Dirk) rides his special three-wheeled bicycle to the rescue of a six-year-old girl (Emily) who has become lost. Because of Dirk's handicaps he is unfairly ridiculed by other boys, but the ridicule doesn't dim his spirit. Dirk's positive attitude, helpful spirit, and unassuming ways all combine to reveal the dignity and value of the disabled. The original music score and lyrics remind viewers of the high value that our Creator places on all of His creation - whether "normal" or "special." Good entertainment, but also valuable in that the story helps viewers to understand and accept the disabilities of others. Includes brief discussion guide. Great for Sunday School or family devotions. *(Also available on 16mm film.)*
©1985

JESUS' LIFE Series - Tape 1: "His Birth"

Content: ★★★★★
Technical Quality: ★★★

PRIMARY AUDIENCE: Ages 5 - Adult
STYLE: Filmation

LENGTH: 35 min.
DISTRIBUTOR: Biblevision
209/825-5645

Improperly titled—this is much more than just a story about the *birth* of Christ. The first in a 7-part series of visualized Scripture reading, this video tells the story of Jesus Christ with the purity of actual word-for-word Scripture (NIV translation). It is unique in that it achieves excellence without cartoons or Hollywood style distortions. While the animation is simple (only "pans and zooms," no character movement), the viewer is drawn into a visual presentation of God's Word that features life-like paintings rather than cartoonish characters. Both the narration and orchestral music and sound effects are superb. This is much more than your ordinary children's video. Because it is broken down into many short stories, it is also excellent for family devotions and even for churches to use as a steppingstone when teaching adult and youth Bible studies on the life and ministry of Christ. This series should be in every home and church video library!
©1986

JESUS' LIFE Series - Tape 2: "His Ministry"

Content: ★★★★★
Technical Quality: ★★★

PRIMARY AUDIENCE: Ages 5 - Adult
STYLE: Filmation

LENGTH: 56 min.
DISTRIBUTOR: Biblevision
209/825-5645

The realistic caricatures, excellent narration, and music enhance this well-produced video which is based upon sections from the NIV Bible. The producer has packed 26 stories of Jesus' ministry onto the 56 minute tape (This is the longest of the seven tapes in the 'Jesus Life' series.) More than entertainment and education, this tape and the series of which it is part is very edifying—honoring God through its visually enhanced Scripture readings. This is the same simple—but effective—animation style as the rest of the series.

©1986

JIMMY AND THE WHITE LIE

Content: ★★★★
Technical Quality: ★★★★

PRIMARY AUDIENCE: Ages 3 - 8
STYLE: Animation

LENGTH: 20 min.
DISTRIBUTOR: Family Films
314/664-7000

This animated adventure is great to share with children at home, church, or school. A baseball-loving young boy accidentally breaks the window of a cranky neighbor and then lies to his parents about what happened. His lie becomes an obnoxious, visible white blob that expands and becomes difficult to hide. Jimmy learns that lies tend to grow and become uncomfortable to live with. He also learns that admitting to a lie is usually much better than being found to be a liar by someone else. Introductory and concluding comments by a real-life animation artist add a valuable, personalized quality.

©1985

JOHN HUS

Content: ★★★★★
Technical Quality: ★★★★

PRIMARY AUDIENCE: Jr. High - Adult
STYLE: Drama

LENGTH: 55 min.
DISTRIBUTOR: Vision Video
215/584-1893

Historically based drama of the beliefs and teachings that led John Hus to be burned at the stake in 1415. He taught that salvation comes by faith, apart from works. He also introduced congregational singing into his church, departing from Latin mass in favor of presenting the Bible in the language of the Bohemian people to whom he preached. His relentless pursuit of God's truth planted the seeds of the Reformation 100 years before Martin Luther. This well-acted and scripted motion picture should be a part of every church and school media library. An inspiring example of one man's commitment to and faith in Jesus Christ. Refusing to compromise his beliefs, he died singing as he was burned at the stake for heresy. This film is not dated in any way. Material will always be "current." (*Also available on 16mm film.*)   ©1977

JOHN WYCLIFFE: The Morning Star

Content: ★★★★
Technical Quality: ★★★★

PRIMARY AUDIENCE: Jr. High - Adult
STYLE: Drama

LENGTH: 1 hr., 15 min.
DISTRIBUTOR: Vision Video
215/584-1893

Fourteenth century scholar and cleric John Wycliffe is best known as the man who first translated the Bible into the English language. However, as this film reveals, Wycliffe also served as a champion of English nationalism against the power of the Pope. Wycliffe once asked "How can men live under the authority of God's Word if they don't even *know* God's Word?" These words ring as clearly in our modern society (where Christians take the availability of God's Word for granted and do not consistently read it) as they did in Wycliffe's own fourteenth century England (where the Bible was not even available in the English language). Very good acting and well-researched historical record, but moves a little slowly throughout. Videos such as this should definitely be integrated into an adult Sunday School curriculum on Church History and made available through every church video lending library! (*Also available on 16mm film.*)   ©1983

JONI

Technical Quality:

Content: ★★★★1/2
★★★★★

PRIMARY AUDIENCE: 10 - Adult
STYLE: Drama

LENGTH: 1 hr., 50 min.
DISTRIBUTOR: World Wide Publications
612/338-3335

Set in picturesque rural New England, this high quality docu-drama opens with a reenactment of the tragic 1967 accident that broke the neck of Joni (Eareckson) Tada. She was instantly transformed from an athletic teenager into a wheelchair-bound quadraplegic. Viewers are drawn into this well-written, technically excellent story as it reveals the physical, emotional, and spiritual struggles that Joni was forced to endure, but always with support from loving family members. JONI asks the questions: "Why does God allow suffering?" and "Why doesn't He always heal those who ask in faith?", but provides no pat answers. Stresses, instead, God's sovereignty, His plan for our lives, and our need to trust Him completely. This is a sensitive drama that provides insights into the emotions and frustrations of what it is like to be paralyzed. Entertaining, inspiring, and challenging. Highly recommended! *(Also available on 16mm film.)*

©1979

JOURNEY OF LIFE

Content: ★★★★1/2
Technical Quality: ★★★★★

PRIMARY AUDIENCE: All ages
STYLE: Documentary

LENGTH: 40 min.
DISTRIBUTOR: Moody Institute of
Science
312/329-2190

Young and old alike will be captivated by the beautiful photography, rich soundtrack, and fast pacing of this Creator-honoring production. It is an eye-opening and awe-inspiring look at how plants have been designed to reproduce and replenish themselves through the incredible voyage of their seedlings. A refreshing and educational examination of some of nature's intricacies, without a single reference to evolution or "millions of years." The movie falls short, however, in that the producer did not use this excellent opportunity to uphold and endorse the Genesis account of creation (specifically, that plants were created "according to their kinds" on Day 3 of the 6 days of creation). Contains a mild evangelistic close that urges viewers to examine God's plan for their lives. Excellent overall and highly recommended. Purchase this one for your video library because one viewing simply is not enough! *(Also available on 16mm film.)*

©1985

Ken Anderson's—
ELIJAH AND THE FIRE FROM HEAVEN

Content: ★★★★
Technical Quality: ★★1/2

PRIMARY AUDIENCE: Ages 4 - 10
STYLE: Animation

LENGTH: 23 min.
DISTRIBUTOR: Ken Anderson Films
219/267-5774

This is one in the "Greatest Stories Ever Told" series of draw-on animation Bible story tapes. It tells and illustrates the account of Elijah, a prophet and man of God. Reinforces the importance of trusting God through times of plenty and times of hardship. Closely follows and further develops the biblical account. Gives interesting details frequently overlooked by other children's Bible story video producers. Elijah is shown receiving food from God's winged messengers (birds) and being nurtured by the widow in the miracle of the meal that never stopped. Children and adults will marvel at the way God used Elijah to bring glory to Himself through contests with the false prophets of Baal and the restored rain in the drought-stricken land. The pace of this episode is good. A timeless, important message that also includes good imagery of Old Testament dress and culture. *(Also available on 16mm film.)* 1982

Ken Anderson's—
JONAH AND THE BIG STORM

Content: ★★★★★
Technical Quality: ★★1/2

PRIMARY AUDIENCE: Ages 4 - 10
STYLE: Animation

LENGTH: 25 min.
DISTRIBUTOR: Ken Anderson Films
219/267-5774

This production does more than just retell the traditional children's version of the story of Jonah and his adventure with the big fish. With respect for communicating the true realities of God's Word—an essential ingredient sadly missing from some of the flashier children's tapes—a fuller story is presented here. Using simple draw-on animation, the producer introduces Jonah, the obedient, God-honoring prophet in his earlier experiences with King Jeroboam (from II Kings 14:25). Then begins the exciting adventure of Jonah trying to run from God and eventually learning how to accept God's grace when people truly change their lives after being warned of imminent judgment. Includes well-woven, interesting information such as how tall the walls around Nineveh were, what a prophet is, and that the king of Nineveh's great-great grandfather was Noah. This is a unique teaching tool that helps transform important biblical truths into applicable lessons for modern Christians. *(Also available on 16mm film.)* 1984

Ken Anderson's—
JOSHUA AND THE PROMISED LAND

Content: ★★★★★
Technical Quality: ★★1/2

PRIMARY AUDIENCE: Ages 4 - 10
STYLE: Animation

LENGTH: 28 min.
DISTRIBUTOR: Ken Anderson Films
219/267-5774

This episode from the "Greatest Stories Ever Told" series uses the life and events of Joshua to teach the importance of trusting God in all circumstances. It shows Joshua as Moses' special helper and eventually as the man that God chose to lead Israel into the Promised Land. A draw-on animated film, it is full of interesting details that help to transform stories which most Christians have probably begun to take for granted—scouting out the land of Canaan, marching around the walls of Jericho, etc.—into an entertaining and genuinely educational format for teaching Old Testament history. This true-to-Scripture series would be excellent for Christian elementary schools to integrate into their Old Testament history curriculums. Also super for VBS, Sunday School, family devotions, and general entertainment. *(Available on 16mm film.)*

1984

Ken Anderson's—
MARY AND JOSEPH

Content: ★★★★1/2
Technical Quality: ★★1/2

PRIMARY AUDIENCE: Ages 4 - 10
STYLE: Animation

LENGTH: 25 min.
DISTRIBUTOR: Ken Anderson Films
219/267-5774

This is an excellent film to show during the Christmas season, but also super as a teaching/entertainment tool throughout the rest of the year, too! The producer has gone to extra effort to allow young viewers to get a glimpse into Jesus' earthly family, being forced to flee to Egypt, the visit by the gift-bearing Magi two years after His birth, being taken to the temple as an infant, and more. Provides details that even mom and dad will appreciate about King Herod, the census, shepherds, the child Jesus teaching in the temple, and other areas. Good to view repeatedly at home, and also as a children's Sunday School audio/visual teaching aid. No apologies for content whatsoever. Only wish the quality of animation was a little better to measure up to the excellent content. *(Also available on 16mm film.)*

1985

Ken Anderson's—
NOAH'S BIG ADVENTURE

Content: ★★★★1/2
Technical Quality: ★★1/2

PRIMARY AUDIENCE: Ages 4 - 10
STYLE: Animation

LENGTH: 23 min.
DISTRIBUTOR: Ken Anderson Films
219/267-5774

This animated adventure takes the exciting story of Noah and the ark out of the realm of fanciful imagination and places it where it belongs as a factual event of history. Opening with a cursory look at the creation of the world, Adam and Eve, and the effects of their sin on the pre-Flood world, it then reveals the story of the world-wide catastrophe/judgment popularly known as "Noah's Flood." Noah's obedience to God in the face of ridicule from his contemporaries is emphasized. This video presents facts generally overlooked or misrepresented in other Noah story tapes—that the animals came to the ark instead of Noah having to go out and search for them, that rain had never before fallen on the earth, and the ark itself is illustrated very similarly to what it probably looked like (a huge barge) rather than a story-bookish little boat with animal heads sticking out all over. Entertaining and wonderfully educational for children. *(Also available on 16mm film.)* 1983

Ken Anderson's—
PAUL'S ADVENTURES

Content: ★★★★★
Technical Quality: ★★1/2

PRIMARY AUDIENCE: Ages 4 - 10
STYLE: Animation

LENGTH: 24 min.
DISTRIBUTOR: Ken Anderson Films
219/267-5774

This video introduces young children to a Bible personality whose adventures and teachings are generally reserved for older audiences. Thanks to this well-researched draw-on animation production, children can now have a visual overview of the life of Paul that will provide increased appreciation during future Sunday School lessons and sermons. Includes many interesting details and integrates simple maps of Israel and the areas visited by Paul on his missionary journeys. Also depicts Paul's conversion, shipwreck, snake bite, beatings, imprisonment, etc. An excellent teaching aid for Sunday School classes, home schools, family devotions, church-time, etc. Closes with challenge to follow Paul's example by sharing the story of Jesus with others. Best suited for children, but teenagers and adults will find this one entertaining and interesting as well! *(Also available on 16mm film.)* 1986

KIDS PRAISE! 4: Singsational Servants

Content: ★★★★
Technical Quality: ★★★1/2

PRIMARY AUDIENCE: Ages 3 - 8
STYLE: Live-action musical

LENGTH: 44 min.
DISTRIBUTOR: Maranatha! Music
714/586-5778

While the graphic design of this video's package may mislead buyers into thinking that the style of presentation is animation, it is really a very good, live-action children's musical, well staged with a colorful set and great costumes. In this story, Charity Churchmouse (seeking fame and fortune on her way to California to become a Gospel music singing star) learns the value of serving others instead of just herself. She finds that serving others puts a song in her heart. If we follow the Lord's example, we'll be servants of all and great in God's kingdom! Constant movement, excellent camera angles, great choreography and a large number of sing-along songs all add to the attention-holding power of this tape. An excellent addition to any video library for the age category listed.

©1984

THE KINGDOM CHUMS: "Little David's Adventure"

Content: ★
Technical Quality: ★★★★★

PRIMARY AUDIENCE: Ages 6 - 10
STYLE: Animation and Live-action

LENGTH: 52 min.
DISTRIBUTOR: Word, Inc.
214/556-1900

Originally telecast during prime time on ABC, "Little David's Adventure" is a combination of high quality animation, live-action drama, great songs and well-executed special effects. It is an excellent tool to begin the process of subtly indoctrinating impressionable young minds with New Age and humanistic religious concepts. In it, three modern day youths are magically changed into cartoon characters after having been transported through a computer screen and a moving rainbow (the "Love Light Beam") to Israel, 3,000 years ago. The children watch as a raccoon (David) slays the giant bulldog (Goliath). (King Saul is a cat and the enemy Philistine army is made up of a variety of ugly rodents.) The raccoon refers both to Jesus and to "the power within," but it is unclear whether the power within is the Holy Spirit or some sort of mystical "force." While this tape admittedly provides 52 minutes of spirited entertainment, it is a poor example of Bible storytelling. Not recommended.

©1986

LAUREL WITH A MORAL

Content: ★★★★★
Technical Quality: ★★★1/2

PRIMARY AUDIENCE: Ages 2 - 5
STYLE: Animation

LENGTH: 32 min.
DISTRIBUTOR: Tyndale Christian Video
708/668-8300

Laurel's animated short stories (there are 12 on this tape) are educational while at the same time thoroughly entertaining. Her stories are told in rhyme and deal with self-esteem, obedience, kindness, generosity, and God's love. You won't mind your children watching this video again and again! An excellent addition to any parent's or grandparent's video library.

mid-1980s

LEARNING ABOUT SEX—
"Where Do Babies Come From?"

Content: ★★★★
Technical Quality: ★★

PRIMARY AUDIENCE: Ages 6 - 8 (and parents)
STYLE: Filmation

LENGTH: 10 min.
DISTRIBUTOR: Family Films
314/664-7000

This simple video tactfully and with just the right amount of information for 6-8 year-olds, answers the question "Where do babies come from?". Best of all, it does so from a Christian point-of-view in the context of our relationship to the God who created us and His plan for reproduction. Reveals the way in which a baby grows inside the mother's womb. Stresses the sanctity of marriage. The *User's Guide* included with the cassette is extremely helpful. This entire video series (of which this is the second segment) is an excellent resource—all of which should be available for families through church and Christian school video libraries.

©1988

LEARNING ABOUT SEX—
"How You Are Changing"

Content: ★★★★
Technical Quality: ★★

PRIMARY AUDIENCE: Ages 8 - 11 (and parents)
STYLE: Filmation

LENGTH: 15 min.
DISTRIBUTOR: Family Films
314/664-7000

The third in the five-part "Learning About Sex" series, this segment is also simply but colorfully animated. HOW YOU ARE CHANGING provides support, encouragement, and accurate, up-to-date medical and physiological information on a level well suited for 8 - 11 year olds. Very interesting and educational. This production is designed to help parents open a channel of communication that will prove to be even more important as the child enters the teen years. As with other segments in the series, sex education is provided from a Christ-centered point-of-view. Helpful *Parent Guide* included.

©1988

LIFE FLIGHT

Content: ★★
Technical Quality: ★★★

PRIMARY AUDIENCE: Jr. High - Adult
STYLE: Drama

LENGTH: 1 hr., 26 min.
DISTRIBUTOR: Vision Video
215/584-1893

The main plot in LIFE FLIGHT is supposed to be that of a Vietnam War veteran's emotional struggle to regain the courage he needs to be able to become part of an emergency medical team which uses a helicopter as a flying ambulance. However, some of the sub-plots (insufficient hospital funds to expand the helicopter program, "do things my own way" attitude of a young helicopter pilot, keep handguns out of the reach of children, etc.) vie with the main theme for preeminence. The viewer is often confused as to the primary message, which appears to change frequently. This story is wholesome, generally well acted and contains some good special effects, but the intensity level associated with these effects somehow never meets the viewer's expectations. On the positive side, one of the best aspects of this film is its challenge to Christians to be knowledgeable about God's Word and to be ready to give an answer to those who ask tough questions. Non-Christians are often interested in knowing a biblical basis for our opinions, a point which is brought out in the movie. There is nothing truly outstanding about this film—too bad since it had potential for much more. *(Also available on 16mm film.)*

©1987

THE LION, THE WITCH, AND THE WARDROBE

Content: ★★★★★
Technical Quality: ★★★★1/2

PRIMARY AUDIENCE: Entire Family
STYLE: Animation

LENGTH: 1 hr., 35 min.
DISTRIBUTOR: Gospel Films
616/773-3361

Awarded with an Emmy as an outstanding animated program, this production of C.S. Lewis's famous novel includes full orchestration. The first tale in Lewis's acclaimed "Chronicles of Narnia," THE LION, THE WITCH, AND THE WARDROBE is the story of two brothers and two sisters who mysteriously pass through a wardrobe closet into the world of Narnia, a land of talking animals and mythical creatures. This story is beautiful and heartwarming in the believability of its symbolism, designed by Lewis to remind us of Jesus' atoning death and resurrection for us and what He has promised He will also do in the future (vanquish Satan and his followers). The kids will plead to view this one again and again, and you will love saying "yes"! HIGHLY recommended! *(Also available on 16mm film.)*   ©1985

THE LITTLE TROLL PRINCE

Content: ★★
Technical Quality: ★★★★★

PRIMARY AUDIENCE: Ages 3 - 10
STYLE: Animation

LENGTH: 46 min.
DISTRIBUTOR: Sparrow Home Video
818/709-6900

This animated Hanna-Barbera production "tells the story of the once-frozen heart of the Little Troll Prince melting into joy upon receiving the greatest Christmas gift of all— God's love." Weak on this point. Should have shown him as actually receiving Christ as Savior rather than being transformed simply by the *realization* that God loves us. "Bu" (the Little Troll Prince) bravely battles the evil ways of the Royal Troll Family and their world where bad is good and no word for love exists. The Royal Troll Family presumably is intended to symbolize Satan and his followers. Excellent with regard to technical aspects, but does not communicate the *true* way to salvation. Better materials dealing with Christmas are available.   ©1987

LITTLE VISITS WITH GOD: Volume 1

Content: ★★★★★
Technical Quality: ★★★1/2

PRIMARY AUDIENCE: Ages 4 - 10
STYLE: Drama, puppetry, animation

LENGTH: 1 hr., 8 min.
DISTRIBUTOR: Family Films
314/664-7000

The vinyl case in which this videocassette is packaged is really a treasure chest in disguise. Ten individual short stories make up this 70 minute tape, and each is a valuable family resource. The great puppetry, real-life dramas, and simple animation segments all combine to reveal God's way to deal with some of life's not-so-smooth situations. Each story covers a specific subject, such as gossiping, being thankful, not looking down on others, how God gives strength, etc. An excellent "family devotions kit," the package also includes a simple *Devotions Guide* that is ideal for either one or two-parent households to lead exciting/interesting family devotions. Each discussion in the *Guide* includes a short memory verse, questions to discuss, suggested short Bible reading, and a suggested prayer. Now there is no excuse not to have at least sporadic times of family devotions! This is one of few Christian films that include Blacks as some of the primary actors.   ©1985

MAKE YOUR OWN MOVIE KIT—
Stories From the Bible

Content: ★★★
Technical Quality: ★★1/2

PRIMARY AUDIENCE: Ages 8 - Adult
STYLE: How-To

LENGTH: 8 min.
DISTRIBUTOR: Sparrow-Star Song
818/709-6900

The video that comes with this kit is but one component of a "complete" outfit. It is very reasonably priced and sure to get your future movie-makers' creative juices flowing. All you need are several friends and a video camera. The kit includes: Hollywood style clapboard and chalk, cardboard director's megaphone, scenery guide, costume guide, audio tape with sound effects, small make-up kit, instructional video (with video title graphics that can be used in your own productions), and three simple scripts. The scripts are "Noah and the Great Flood," "The Good Samaritan," and "The Story of Jesus Birth." A great "TV Alternative" in itself, producing the three stories included in this kit would also make a fun, interactive, and truly memorable special project for Sunday School, youth groups, family devotions, etc.   ©1988

MARTIN LUTHER

Content: ★★★★1/2
Technical Quality: ★★★★

PRIMARY AUDIENCE: Jr. High - Adult
STYLE: Drama

LENGTH: 1 hr., 45 min.
DISTRIBUTOR: Vision Video
215/584-1893

This dramatic presentation of the life of Martin Luther was originally released in theaters worldwide in the 1950s. Although it is in black and white and moves along a little slower than more recent films, it is technically excellent in almost every respect and was nominated for an Academy Award. Documents Luther's uncompromising convictions and intense, lifelong questioning of the Catholic church's fund raising techniques and traditionally accepted interpretations of Scripture. Should be a staple component of every church and school video lending library because it provides an excellent historical perspective of the birth of Protestant denominations from their pre-Reformation "parent," Roman Catholicism. *(Also available on 16mm film.)*

1950s

MARTIN LUTHER - HERETIC

Content: ★★★★★
Technical Quality: ★★★★★

PRIMARY AUDIENCE: Jr. High - Adult
STYLE: Drama

LENGTH: 1 hr., 8 min.
DISTRIBUTOR: Family Films
314/664-7000

This production is top quality in every respect—content, acting, cinematography, etc. Very effectively communicates the historical facts and theological upheaval that surrounded Luther during the Reformation. We witness his quest for truth as it leads him to discover the doctrine of "salvation through grace alone"—a concept in direct conflict with his Catholic church and the state! This production challenges its viewers to review their own convictions, to strengthen their commitment to the same Lord that Luther so deeply loved. A wonderful church history film for every church or Christian school video lending library. *(Also available on 16mm film.)*   ©1981

McGEE AND ME!—
"The Big Lie" (Episode 1)

Content: ★★★★★
Technical Quality: ★★★★★

PRIMARY AUDIENCE: Ages 5 - 12
STYLE: Drama with animation

LENGTH: 28 min.
DISTRIBUTOR: Tyndale Christian Video
708/668-8300

This well-written story uses both animation and drama. The message is clear: even what you may believe to be just a harmless little fib can grow and grow until it reveals itself as a very harmful lie. The film shows clearly that lies hurt not only the one whom the lie is directed against, but also the one doing the lying. Nicholas (a new boy in the neighborhood) and McGee (a cartoon character who gives Nicholas bad advice) discover that telling lies is a sure way to hurt others. Nicholas bows to peer pressure and spreads a fanciful story about an elderly Indian man who lives alone in his neighborhood. As a result, the school bully and his gang vandalize the home of the Indian man. Nicholas apologizes to the man for the damage his lie has caused and then acts on his apology and remorse by cleaning up the massive mess created by the gang. The portrayal of a Christian father is sensitive and an example of a godly head of household. The music track is upbeat and contemporary but not overpowering. A *Discussion Guide* for parents accompanies each video in this series. *(Also available on 16mm film.)*
©1989

McGEE AND ME!—
"A Star in the Breaking" (Episode 2)

Content: ★★★★
Technical Quality: ★★★★★

PRIMARY AUDIENCE: Ages 5 - 12
STYLE: Drama with animation

LENGTH: 28 min.
DISTRIBUTOR: Tyndale Christian Video
708/668-8300

A STAR IN THE BREAKING shows young Nicholas enjoying a brief opportunity at stardom, and it is his entertaining role that reminds us of the biblical warning that "pride comes before a fall." Nicholas's name is selected as a contestant in the popular children's TV game show "Trash TV." Word spreads through his school of the upcoming TV appearance and Nick is an instant celebrity. However, his sudden notoriety goes to his head, and he quickly becomes very proud and self-centered. On the TV game show, Nicholas miserably loses an extremely messy food fight to the other contestant, a girl! Humiliated, the show ends with Nicholas recognizing the error of his prideful attitude and viewers witness his mother (who has been humbled herself during the episode) telling him that "God gives grace to the humble, but He flattens the hotshot." *(Also available on 16mm film.)*
©1989

McGEE AND ME!—
"The Not-So-Great Escape" (Episode 3)

Content: ★★★★★
Technical Quality: ★★★★★

PRIMARY AUDIENCE: Ages 5 - 12
STYLE: Drama with animation

LENGTH: 28 min.
DISTRIBUTOR: Tyndale Christian Video
708/668-8300

This episode opens with an animated contemporary parable and then switches to a combination of live-action and integrated animation. THE NOT-SO-GREAT ESCAPE is fast paced, humorous, and intense. It imparts excellent biblical principles by stressing the value of obedience, being wise when making decisions, and carefully guarding what is allowed to enter our minds. The young star of the series, Nicholas, sneaks out of his house to view a forbidden movie with a friend. Nicholas finds the horror film to be grotesque and revolting. At home, his disappointed mother and father sternly but lovingly confront him with his disobedient actions and explain their desire to protect him from detrimental images that become permanently impressed on his mind. He is given an appropriate punishment. Parents, especially the father, are very well handled. This tape should be a permanent component of any video library. *(Also available on 16mm film.)* ©1989

McGEE AND ME!—
"Skate Expectations" (Episode 4)

Content: ★★★★★
Technical Quality: ★★★★★

PRIMARY AUDIENCE: Ages 5 - 12
STYLE: Drama with animation

LENGTH: 27 min.
DISTRIBUTOR: Tyndale Christian Video
708/668-8300

This fourth episode of the McGEE AND ME! series contains something that parents have long been wishing for: an eleven-year-old role model whose selfless actions are based on a desire to live what the Bible teaches! The tape opens with a very humorous animated adaptation of the Luke 10 story of the Good Samaritan. Then, Nicholas (the young star of the series) is presented with an opportunity to be a modern-day good Samaritan when he comes repeatedly to the aid of a younger boy who is being taken advantage of by the school bully. Exciting skateboard sequences, upbeat, tasteful music, and Nicholas's animated sidekick "McGee" all combine to make this a truly top quality show. Teaches that cheating doesn't pay while encouraging viewers to look for ways that they can be Good Samaritans themselves. *(Also available on 16mm film.)* ©1989

McGEE AND ME!—
"Twister and Shout" (Episode 5)

Content: ★★★★★
Technical Quality: ★★★★★

PRIMARY AUDIENCE: Ages 5 - 12
STYLE: Drama with animation

LENGTH: 25 min.
DISTRIBUTOR: Tyndale Christian Video
708/668-8300

In this episode, Nicholas learns that when your trust is in God, you don't have to worry about being alone. Mom and dad have gone out for the evening and left "big sister" in charge of "the kids." During mom and dad's absence, a tornado cuts off all electrical power and the phones go dead. A tree limb even crashes through the living room window. This small group of squabbling, sarcastic kids learns that God has given them to one another to help each other in times of need. Big sister learns that she isn't quite as self-sufficient as she thought. A top quality drama with excellent animation sequences interspersed throughout. Great music soundtrack. *(Also available on 16mm film.)*
©1989

MIKE WARNKE—Do You Hear Me?!

Content: ★★★1/2
Technical Quality: ★★★★

PRIMARY AUDIENCE: Jr. High - College
STYLE: Live comedy

LENGTH: 1 hr., 6 min.
DISTRIBUTOR: Word, Inc.
214/556-1900

To those who appreciate his sense of humor, Mike Warnke is a lovable, middle-aged teenager who happens to have long hair and wears an earring. The quality of his energetic comedy routine is equal to almost any late night TV comedian only better, because this born-again former Satanist will have you laughing uproariously without telling a single dirty joke or using even one word of profanity. Recorded live in a large church auditorium, Warnke's first few minutes are spent defending his own outward appearance, promptly followed by jokes about those who judge the Christian commitment of others based solely on physical appearances. This belabored point is well taken, but may cause more cases of teenage rebellion than the removal of judgmental attitudes. Although it is not entirely clear, the main point of Warnke's message seems to be to challenge Christians to get out of their traditional "comfort zone Christianity" and to put their verbalized beliefs into actions. ©1989

THE MORMON DILEMMA

Content: ★★★★
Technical Quality: ★★

PRIMARY AUDIENCE: High School - Adult
STYLE: Drama

LENGTH: 1 hr., 7 min.
DISTRIBUTOR: Bridgestone Video
619/431-9888

While this low-budget drama leaves a bit to be desired in terms of technical and acting quality, it none-the-less deserves to be made a part of every church video library. THE MORMON DILEMMA shows a young Mormon couple confronting their neighbor (Jim) because the Christian church that he attends has just shown a film that they understand to be an attack against their faith. During an exchange later that evening, both the Mormon couple and Jim discover the deep differences between Mormonism and evangelical Christianity. If you have Mormon friends or acquaintances, you need to view this video—repeatedly! An entertaining and eye-opening attempt to educate both Christians and Mormons.   ©1988

MUSIC BOX

Content: ★★★
Technical Quality: ★★★★

PRIMARY AUDIENCE: Jr. High - College
STYLE: Drama

LENGTH: 28 min.
DISTRIBUTOR: Word, Inc.
214/556-1900

If you are in the mood for some lighthearted, silly entertainment designed to put joy back into the Christian life, see this production. This is probably the most creative, upbeat representation of "ministering spirits" ever filmed. In this story a factory worker whose life is filled with drudgery is visited by five dancing, singing, white tuxedo-clad angels. They leave him a gift—a music box—which brings great joy and totally changes the factory worker's attitude toward life. Afraid to share the gift with others, he is again visited by the angels and learns that the gift must be shared and not hoarded or kept for oneself. The parabolic message—to share the joy-filled message of salvation—is great. However, these frolicking angels border on, and undoubtedly in many minds cross the border into, sacrilegiousness. Don't watch this one unless you're prepared to smile! *(Also available on 16mm film.)*   ©1980

ORDINARY GUY

Content: ★★★★1/2
Technical Quality: ★★★★

PRIMARY AUDIENCE: Entire family
STYLE: Drama

LENGTH: 1 hr., 10 min.
DISTRIBUTOR: Daystar Productions
708/541-3547

Delightfully humorous, this award-winning drama has proven itself as a classic among Christian film productions. The message is timeless—one which challenges "ordinary" Christians to be rich in good works and share the message of God's love. ORDINARY GUY shows how the Lord can use all of us, even if we're not a famous athlete, musician, or politician. It is a well-written reflection of real life. In terms of the message communicated, this has got to be one of the top Christian family films of all time!! Buy this one for your church's video lending library. NOTE: Includes very short, tactfully handled scenes of gang violence. *(Also available on 16mm film.)* ©1979

ORIGINS: How The World Came to Be

Content: ★★★★★
Technical Quality: ★★★★★

PRIMARY AUDIENCE: High School - Adult
STYLE: Documentary

LENGTH: 6 episodes, approx. 30 min.
each
DISTRIBUTOR: Films For Christ
602/894-1300

The most thorough yet easy-to-understand films ever produced on the Creation/ evolution controversy, this award-winning documentary series consists of six episodes. The titles are: *The Origin of the Universe, The Earth—A Young Planet?, The Origin of Life, The Origin of Species, The Origin of Mankind,* and *The Fossil Record.* Moderated by internationally respected Christian scientist Dr. A.E. Wilder-Smith, the production style is very similar to that of a National Geographic Special—but from a six-day scientific creationist rather than an evolutionary perspective. Refreshing, attention-holding variety of visuals include nature photography, animation, interviews, dramatizations, etc. shot in many settings throughout the world. The evolution theory has turned thousands away from faith in Christ. The ORIGINS series should be a staple of every church's video library because it provides a tremendous defense for the trustworthiness and infallibility of God's Word. Informative and awe-inspiring. *(Also available on 16mm film.)*
©1982

OUR DWELLING PLACE Series—
"The Trials of Jesus"

Content: ★★★★
Technical Quality: ★★1/2

PRIMARY AUDIENCE: Ages 4 - 7
STYLE: Animation

LENGTH: 24 min.
DISTRIBUTOR: Rainfall, Inc.
616/245-5985

This biblically accurate video is part of the OUR DWELLING PLACE series and contains four short story segments: "The Last Supper," "The Garden of Gethsemane," "The Crucifixion," and "Easter Appearances." Each segment is only a few minutes long and is separated by brief cut-backs to animated conversations between several modern day orphan children. Good possibility as an enhancement for family devotions. Very little character movement; the story segments consist mainly of pans, zooms and dissolves between various rough illustrations. Animation is simple but quick editing helps to keep the viewer's attention. ©1988

THE PARABLES: Volume 2

Content: ★★★★★
Technical Quality: ★★★

PRIMARY AUDIENCE: Ages 3 - 8
STYLE: Animation

LENGTH: 25 min.
DISTRIBUTOR: Tyndale Christian Video
708/668-8300

Eleven short cartoons, each with a different message, make up this volume of the PARABLES series. Each cartoon was originally broadcast in Canada and the USA as a part of the "Circle-Square" Christian children's television programs. The quality of animation is good, consisting of rather simple yet colorful full-movement animals, people, and objects. The moral of each short segment (1 to 4 minutes each) is excellent and always biblically based. The brevity of each cartoon makes them perfect as contemporary visual illustrations of Bible parables for family devotions. You'll be glad to let your children view this tape over and over again! ©1983

PARABLES FROM NATURE: Series 5

Content: ★★★★★
Technical Quality: ★★★1/2

PRIMARY AUDIENCE: Ages 3 - 8
STYLE: Videostrip

LENGTH: 32 min.
DISTRIBUTOR: Cathedral Films
818/991-3290

This series is an upgraded video edition of some excellent Sunday School filmstrips that many parents will recognize from their own growing-up years. Because this is a "videostrip," it does not contain camera or character movement of any kind. However, the colorful illustrations are top quality and the use of friendly animals communicates to preschool and early elementary-aged children quite well. There are three animal "parables" on every video. This tape, #5, contains "Peppy the Pup"/The Prodigal Son, "Chuckie Chipmunk"/The Good Samaritan, and "Bushy the Squirrel"/ The Foolish Rich Man. Each animal parable is immediately linked to life-like illustrations and a quick summary of Jesus' original parable. Excellent not only for Sunday School, but also for family devotions and as wholesome entertainment communicating biblical principles. Includes a mini-Study Guide to assist parents/ teachers in leading follow-up discussions if desired.   ©1987

PEDRO AND THE BRIGHT CITY

Content: ★★★★
Technical Quality: ★★

PRIMARY AUDIENCE: Ages 3 - 7
STYLE: Videostrip

LENGTH: 13 min.
DISTRIBUTOR: Child Evangelism
Fellowship
314/456-4321

This video is a remake of a story that was originally produced in the 1950s on filmstrip. By "panning" left and right and "zooming" in and out from colorful but non-moving illustrations, this timeless story of Pedro, a young Peruvian mountain boy, still captures the attention of modern children as completely as it did their parents a full generation earlier. In the story, Pedro steals a toy alpaca (horselike animal), narrowly escapes death in a fall over a cliff, and later learns from a friendly missionary who has a copy of CEF's "Wordless Book" that the Lord forgives sin. The gospel message is well presented, making this a good audio-visual tool for VBS, children's Sunday School, church time, youth groups, etc.   ©1956, 1984

PERFECT PEACE

Content: ★★★★★
Technical Quality: ★★★★★

PRIMARY AUDIENCE: All Ages
STYLE: Live action

LENGTH: 28 min.
DISTRIBUTOR: Moody Institute of
Science
312/329-2190

Another in the series of "praise and worship" videos produced by Moody in cooperation with Maranatha! Music, this video combines beautiful songs, breathtaking nature photography, and moving passages from God's Word. Top-notch technical aspects include footage of thousands of beautiful Monarch butterflies, large variety of animals in their natural habitats, fast-moving cloud formations enhanced with time lapse photography, and a kayak running dangerous river rapids. Extremely interesting and awe-inspiring for adults, but also sure to captivate and relax even the rowdiest youngster! The stereo soundtrack is tremendous if shown on a stereo VCR and TV.

©1989

PILGRIM'S PROGRESS

Content: ★★★★
Technical Quality: ★★★

PRIMARY AUDIENCE: Ages 10 - Adult
STYLE: Animation

LENGTH: 38 min.
DISTRIBUTOR: Gospel Films
616/773-3361

In the 1950s, pioneer Christian filmmaker C.O. Baptista produced a colorful, animated version of John Bunyan's classic novel, PILGRIM'S PROGRESS. In the mid-1970s, the production was updated with the addition of contemporary Christian music written and performed by Chuck Girard. The novel is an excellent allegory of the Christian life both as it should be lived and as it is generally lived. This visual abbreviation of the book does a good job of illustrating the fact that prayer and the armor of God (Ephesians 6) are necessary to stand against the fiery darts of Satan throughout our Christian "pilgrimage." The animation is lifelike and with nice detail, but transitions between various segments of the story are often abrupt. This particular production has a lower intensity level than DANGEROUS JOURNEY (another animated version of Bunyan's novel) so is probably better suited for young children. *(Also available on 16mm film.)*

©1976

THE PRODIGAL

Content: ★★★★1/2
Technical Quality: ★★★★★

PRIMARY AUDIENCE: Ages 10 - Adult
STYLE: Drama

LENGTH: 1 hr., 46 min.
DISTRIBUTOR: World Wide Publications
612/338-3335

THE PRODIGAL is a top-quality drama based upon a well-written, realistic script. It is the story of Greg, a talented 21-year-old tennis player, who is under so much pressure from his highly successful businessman father to conform to the father's expectations that emotionally charged sparks inevitably fly whenever the two are together. At one point Greg lashes out against his father's worldly, hypocritical brand of Christianity. The intense honesty of this late-night confrontation causes the father (and all Christian viewers) to begin to examine seriously the true depth of his relationship with Christ. While the film moves a little slowly during the first 20 minutes, it soon captures all viewers for the duration. Excellent screenplay, excellent acting, excellent in all technical aspects. Sponsor this one for your church's video lending library! *(Also available on 16mm film.)*   ©1983

THE PRODIGAL PLANET

Content: ★★★1/2
Technical Quality: ★★★★

PRIMARY AUDIENCE: Jr. High - Adult
STYLE: Drama

LENGTH: 2 hrs., 7 min.
DISTRIBUTOR: Mark IV Pictures
515/278-4737

THE PRODIGAL PLANET is the fourth in the dramatic prophecy series produced by Mark IV (other episodes are A THIEF IN THE NIGHT, A DISTANT THUNDER, and THE IMAGE OF THE BEAST). This is a futuristic film that attempts to portray some of the true events that will take place during the seven-year reign of the anti-Christ on earth. This film depicts a world devastated by nuclear holocaust toward the end of the tribulation period. Seeing the seven-year tribulation lived out in the three final films of this series throws a powerful light on what is yet to come. Definitely not exactly as things will be, but probably similar in many respects. Answers some of the basic questions of the Christian faith through dialogue during the film. It's greatest strength lies in its usefulness as an evangelistic tool. Greatest weakness is its length. Intriguing introduction and very good special effects. *(Also available on 16mm film.)*
©1983

QUESTIONS PARENTS ASK - Dr. James Dobson

Content: ★★★★
Technical Quality: ★★

PRIMARY AUDIENCE: Parents and prospective parents

STYLE: Interview

LENGTH: 58 min.
DISTRIBUTOR: Word, Inc.
214/556-1900

Most parents will appreciate the humorous, down-to-earth responses of Dr. Dobson to the questions of a live TV-studio audience. Originally recorded and distributed as part of a low budget, multi-segment TV series, the approach is reminiscent of early "Donahue" shows, but without all the tension and negativism. Includes three separate 20 minute tapings: 1) "The World of Children," 2) "Self-Esteem in Children," 3) Coping with Toddlerhood." Somewhat dated by clothing styles, but the content is practically timeless. A good addition to your church's video lending library.   ©1982

QUIGLEY'S VILLAGE: "Be Kind to One Another"

Content: ★★★★★
Technical Quality: ★★★★

PRIMARY AUDIENCE: Ages 3 - 8
STYLE: Puppets and live-action

LENGTH: 36 min.
DISTRIBUTOR: Chariot Video
708/741-2400

Each segment in the QUIGLEY'S VILLAGE series combines live actors together with puppets, very similar in style to SESAME STREET® but with a biblically-based approach to communicating values. Mr. Quigley is a very friendly man whom the animal puppets rely upon for sound advice. In this episode, it's time for the "World Championship Hide-and-Seek Contest" and Danny Lion and Spike the Porcupine both want to be the champ. When Danny breaks his arm, Spike thinks it serves him right for thinking that she won't win the championship. But when Danny needs Spike's help, they both learn a little more about what being kind really means. From the examples of the puppets and the good advice given by Mr. Quigley, children will understand why it is important to be kind to one another. Contains three very good sing-along songs that every parent will be happy to hear their children singing throughout the day. Well-written (age appropriate) script, excellent original music, good quality sets, and very good visual effects.   ©1987

QUIGLEY'S VILLAGE: "Fun Aerobics for Kids!"

Content: ★★★★
Technical Quality: ★★★★

PRIMARY AUDIENCE: Ages 3 - 8
STYLE: Puppets and live-action

LENGTH: 22 min.
DISTRIBUTOR: Chariot Video
708/741-2400

This children's exercise video is an excellent solution to the problem of "couch potato" kids. Tie on the tennis shoes, insert the tape, and get out of one another's way! Integrating the people and puppets from the QUIGLEY'S VILLAGE series, and adding the talents of certified aerobics instructor Cheryl Merrill, "Fun Aerobics for Kids!" is both entertaining and energetic. Two adults and several children do various creative exercises, leading viewers in an aerobic workout. A good use of video special effects and Christian children's music combine to create an exercise routine that even mom will want to join! Parents will also appreciate the fact that no skimpy or expensive aerobics outfits are used. Children wear normal, loose-fitting shirts along with shorts and sneakers. A good video to include as a regular part of their exercise schedule, especially on rainy or snowy days.   ©1988

THE ROLE OF PRAYER IN SPIRITUAL AWAKENING

Content: ★★★★★
Technical Quality: ★1/2

PRIMARY AUDIENCE: High School - Adult
STYLE: Lecture

LENGTH: 26 min.
DISTRIBUTOR: Inspirational Media
714/886-5224

The message is definitely the strong point of this tape. The speaker, Dr. J. Edwin Orr, reveals through several amazing, factual stories that all the great spiritual awakenings (revivals) of the past 200 years began as a result of Christians uniting together for prayer. He opens with a story about the moral climate of the American people at the time of the Revolutionary War and then proceeds to share many other equally startling accounts from both sides of the Pacific Ocean. While this is an important message for every Christian to hear, it is not really necessary to view. The barely acceptable production standards do not enhance Dr. Orr's relaxed, non-animated style of speaking. An audio cassette of this valuable message would probably be just as effective as the video (if an audio cassette were available).   ©1977

SAMMY

Content: ★★★★
Technical Quality: ★★★

PRIMARY AUDIENCE: Entire family
STYLE: Drama

LENGTH: 1 hr., 8 min.
DISTRIBUTOR: Mark IV Pictures
515/278-4737

This is a heartwarming story that explores the tensions, conflicts, and triumphs of a young crippled boy and his financially stressed family. This award-winning movie will appeal to parents and children alike. A variety of animals, especially one particular kitten, add a soft touch to the story and really hold the younger viewers' attention. The message and story are good overall. *(Also available on 16mm film.)* ©1977

SANDI PATTI: "Make His Praise Glorious"

Content: ★★★★★
Technical Quality: ★★★★

PRIMARY AUDIENCE: Adults
STYLE: Live Concert

LENGTH: 1 hr., 59 min.
DISTRIBUTOR: Word, Inc.
214/556-1900

If you can't actually attend a Sand Patti concert, this is the next-best thing. Recorded before an audience of thousands, this video contains a wide variety of camera angles, lighting effects, etc., to capture the energy and beauty of Sandi's concert. The emphasis of the songs is on praise. Get a sitter or put the kids to bed, and then take the phone off the hook, turn down the lights, and insert this tape for a two hour "date" in your own living room. Make your evening all-the-more special by dressing up as if you were actually going out to the concert, then sit back and cuddle! ©1989

SATANISM UNMASKED

Content: ★★★★
Technical Quality: ★★1/2

PRIMARY AUDIENCE: High School - College
STYLE: Documentary

LENGTH: 1 hr., 48 min.
DISTRIBUTOR: Infinity Video
918/582-2126

SATANISM UNMASKED is a startling, humorous, and sometimes frightening revelation of modern Satanism in the USA. This video has been produced in a way that is guaranteed to capture and hold the attention of its intended teenage audience, but which is also guaranteed to "turn off" many adults. Opens with an excellent attention-grabbing black and white music video. Uses high energy contemporary Christian music, interviews with famous former Satanists (Johanna Michaelsen, Lauren Stratford, Sean Sellers), and several silly but message-carrying skits to emphasize the fact that Satan is alive and active throughout America, but that born-again Christians have absolute power over him and his demons through our position in Jesus Christ. Makes an attempt at instructing viewers in how to exercise our authority but is weak on this point. Consider watching this show as a group and then discussing it afterward to more fully comprehend and enhance what has been seen.

©1988

SHIOKARI PASS

Content: ★★★★★
Technical Quality: ★★★

PRIMARY AUDIENCE: High School - Adult
STYLE: Drama

LENGTH: 1 hr., 3 min.
DISTRIBUTOR: World Wide Pictures
612/338/3335

There are very few translated films that American audiences are willing to watch. Lip-sync usually isn't perfect, so, unless it's "King Kong vs. Godzilla" we tend to think that watching a translated film is not worthy of our TV time. SHIOKARI PASS is a wonderful exception to this false standard. Set in turn-of-the-century Japan, this true story centers around a young Japanese man who became a Christian and, though frequently taunted, proved that his Christian faith really made a difference when he sacrifices his own life to save dozens of train passengers from certain death. *(Also available on 16mm film.)*

©1978

SPACE SHUTTLE JOURNEY

Content: ★★★★1/2
Technical Quality: ★★★★

PRIMARY AUDIENCE: Ages 3 - 10
STYLE: Animation

LENGTH: 25 min.
DISTRIBUTOR: Ken Anderson Films
219/267-5774

This Christian cartoon is one that youngsters ask to see over and over again. The basic story is that of two children and their fathers who go on a Space Shuttle flight to conduct a "comet search" experiment which the boys developed as a school project. While this is not too realistic, it certainly captures the imaginations of children and provides an exciting story into which the producer has woven an important biblical lesson: Don't be disappointed if things don't seem to go the way you want them to go because God is in control. He knows what is best for you. A winner for any family or church video library. *(Also available on 16mm film.)*   ©1985

A SPORTS ODYSSEY

Content: ★★★
Technical Quality: ★★★★

PRIMARY AUDIENCE: Ages 10 - Adult
STYLE: Documentary

LENGTH: 40 min.
DISTRIBUTOR: Bridgestone Group
619/431-9888

Excellent for those who already have a love for sports but is also exciting enough that even non-sports enthusiasts will enjoy it. This is the type of show that will be viewed over and over again. A good video to watch when relatives are over for the winter holidays, especially for men and teens. Content is heavy on surfing and skiing clips. Not much of baseball, football, or basketball on this one. Testimonials toward conclusion are concise and non-preachy. Good! *(Also available on 16mm film.)*
©1976

STORMIE OMARTIAN'S First Step Workout Video

Content: ★★★★★
Technical Quality: ★★★★

PRIMARY AUDIENCE: Ages 10 - Adult
STYLE: How-To

LENGTH: 32 min.
DISTRIBUTOR: Sparrow Home Video
818/709-6900

This "workout video" is excellent for beginners or for anyone whose physical condition prohibits them in one way or another from strenuous exercise. Low impact, low stress, low sweat, but very high quality and non-intimidating. Features singer/ songwriter Stormie Omartian with several smiling friends, all of whom are conservatively dressed (in terms of workout attire, that is). Video builds from tension relaxing exercises to toning and stretching warm-up exercises to seven minutes of low impact aerobics, then gradually works the exerciser back down through firming and strengthening exercises, and then through final stretching exercises. No complicated exercise maneuvers. If you can walk and move your arms at the same time, you can benefit from this tape. Workout accompanied by beautiful instrumental renditions of Christian music.
©1988

STORMIE OMARTIAN'S Low Impact Aerobic Workout

Content: ★★★★★
Technical Quality: ★★★★1/2

PRIMARY AUDIENCE: Ages 10 - Adult
STYLE: How-To

LENGTH: 33 min.
DISTRIBUTOR: Sparrow Home Video
818/709-6900

Be sure to clear a generous space in front of the TV before you slip on your sweats, tie on your tennis shoes, and push the video "play" button. This excellent low impact aerobic workout makes high energy exercise almost fun as you workout to the instrumental sounds of many contemporary Christian songs. Stormie is joined throughout the exercise routine by four other women (including Charlene Tilton of "Dallas" fame) on a simple but artistic studio set. An excellent video to give to someone you know who is serious about getting back into or maintaining her physical condition.
©1987

SUNSHINE FACTORY: Anger

Content: ★★★
Technical Quality: ★★★1/2

PRIMARY AUDIENCE: Ages 4 - 10
STYLE: Live action

LENGTH: 23 min.
DISTRIBUTOR: Word, Inc.
214/556-1900

This program combines adult and children as actors and actresses along with puppets, simple video effects, and lively message-based songs to weave a good story about anger. "Mr. Crabtree" (the slovenly villain) discovers that if he can make people angry, he can slow down the Sunshine Factory. He and his puppet sidekick "Smudge" temporarily succeed in their attempts. Eventually, however, "P.J." (the fix-it shop repairman) gets the gang to smiling and singing together, which in turn gets the Sunshine Factory running again. This is a morally positive story but would be better suited for use by Christian families if a Bible-based explanation of why anger can be detrimental were interwoven into the message. ©1984

SUPERBOOK: Volume 5

Content: ★1/2
Technical Quality: ★★★1/2

PRIMARY AUDIENCE: Ages 5 - 10
STYLE: Animation

LENGTH: 40 min.
DISTRIBUTOR: Tyndale Christian Video
708/668-8300

The "Superbook" series is another in the genre of animated Bible stories that, unfortunately, inject contemporary children into each episode. The problem with this type of storytelling is that young viewers (those 5 to 7 years of age and younger) cannot readily separate fiction from reality. Studies indicate that they soon come to believe that the fictional characters are as much a part of the scriptural record as the Bible characters themselves, or they assume that because the children's part is fictional, the Bible story must be fiction, too. Although "Superbook" episodes are very entertaining and excellent attention holders, younger viewers would be better off watching animated Bible stories such as Ken Anderson's "Draw-On" series instead. This tape—one of dozens available in this series—contains THE TEST (Abraham's willingness to sacrifice Isaac) and HERE COMES THE BRIDE (Abraham sends his servant to find a wife for Isaac). Both episodes contain more scriptural inaccuracies and oversights than a reasonable amount of "artistic license" should allow. Read the story to your children directly from the Bible rather than expose them to these poorly researched representations of important historical events. ©1982

SUPERCHRISTIAN

Content: ★★★★1/2
Technical Quality: ★★★

PRIMARY AUDIENCE: Jr. High - College
STYLE: Drama

LENGTH: 28 min.
DISTRIBUTOR: Gospel Films
616/773-3361

Awarded "Best Christian Youth Film" of 1980, SUPERCHRISTIAN contains just the right balance of humor and seriousness to be entertaining and challenging. This drama allows Christians both to laugh at themselves and to take an introspective look at their own relationship to Christ. The story of this film raises the question, "Are you an actor playing the role of a Christian, or are you a servant of the Most High God?" The star of the show, Clark Cant, throws away his "Christian costume" after being confronted by another brother in the Lord who also used to act Christian on Sundays but lived like the rest of the world throughout the other six days of the week. Especially well suited for Jr. High through college youth, but a message that Christian adults would do well to see, too! A good title to add to your church's lending library. *(Also available on 16mm film.)*

©1980

SWISS FAMILY ROBINSON: Volume 1
"The Terrible Typhoon"

Content: ★★★★
Technical Quality: ★★★1/2

PRIMARY AUDIENCE: Entire family
STYLE: Animation

LENGTH: 25 min.
DISTRIBUTOR: Chariot Video
708/741-2400

A dramatic cartoon series that will be enjoyed and appreciated by the entire family, mom and dad, too! THE TERRIBLE TYPHOON is the first of many episodes in this rather substantially revised version of the classic novel, *Swiss Family Robinson*. In this segment, the viewer is introduced to the Robinson family while they are aboard a luxury sailing ship headed for Australia. They are a loving, praying family with three children (two boys and a girl). Several humorous and heartwarming scenes lead up to the sudden eruption of a fierce storm that lasts for several days and finally causes the ship to run aground on a reef. Four of five members of the Robinson family are left stranded without a life raft aboard the shipwrecked vessel while the fifth member of the family has been blown overboard by the howling typhoon. This segment ends at this point.

©1988

SWISS FAMILY ROBINSON: Volume 3
"Mysterious Island"

Content: ★★★★
Technical Quality: ★★★1/2

PRIMARY AUDIENCE: Entire family
STYLE: Animation

LENGTH: 25 min.
DISTRIBUTOR: Chariot Video
708/741-2400

The Robinson family has built a makeshift raft which they are sailing to a nearby deserted island. They watch in horror as their once-trusted luxury vessel slips further into the sea. Surviving some pretty tense moments in their search for a beach upon which to land their raft, they are finally able to go ashore. The family realizes that they are now stranded with no way home. Also, apparently, no other passengers or crewmen have survived. However, they are thankful to God for sparing them. Segment ends with them welcoming an orphaned baby animal into their family. Successive episodes show the building of their treehouse and how they cooperate as a family in reliance upon God to survive and flourish on this deserted Island. This cartoon series will undoubtedly be viewed again and again and is appropriate for almost every age group.   ©1988

A THIEF IN THE NIGHT

Content: ★★★★
Technical Quality: ★★★1/2

PRIMARY AUDIENCE: Jr. High - Adult
STYLE: Drama

LENGTH: 1 hr., 9 min.
DISTRIBUTOR: Mark IV Pictures
515/278-4737

This classic film was the first and continues to be the most popular of what eventually became a four-part dramatic prophecy series written from a pre-tribulational rapture point-of-view. The other three titles are: A DISTANT THUNDER, THE IMAGE OF THE BEAST, and PRODIGAL PLANET. This particular production weaves the story of a young woman (Patty) who awakes one morning to find that her husband and millions of other Christians have all simultaneously disappeared from the face of the earth. "As dramatic, earthshaking events begin to unfold around her, Patty realizes she is living in the end times spoken of in biblical prophecy." Concludes with a thought-provoking cliffhanger that serves well to end this film and to provide a bridge into the next (A DISTANT THUNDER). While the film is dated by early 1970s clothing and hair styles, the story is generally well written and acceptably acted overall. Clearly presents the way of salvation. Entertaining, evangelistic, and educational all at the same time. *(Also available on 16mm film.)*   ©1972

THIS IS THE DAY

Content: ★★★★★
Technical Quality: ★★★★★

PRIMARY AUDIENCE: All ages
STYLE: Live Action

LENGTH: 28 min.
DISTRIBUTOR: Moody Institute of
Science
312/329-2190

By combining a beautiful collage of some of the most awe-inspiring nature photography and praise and worship music ever assembled, this video is sure to instill in its viewers a deep respect and appreciation for the God of creation. THIS IS THE DAY contains "classic hymns rendered in a warm, contemporary style" and is accompanied by several moving passages from Scripture, each beautifully complementing the scenes (cascading waterfalls, gorgeous sunsets, starry skies, beautiful birds) and sounds that make this a relaxing look at the power, majesty, and promises of God. This video is a great complement to personal devotions or family worship, or to put on any time you need a relaxing mood. (A sure hit with young mothers!)   ©1988

TREASURE CHEST Series:
"Zacchaeus - Little Man up a Tree"

Content: ★★1/2
Technical Quality: ★

PRIMARY AUDIENCE: Ages 2 - 5
STYLE: Drama

LENGTH: 14 min.
DISTRIBUTOR: Christian Duplications,
Inc.
407/299-7363

You can't judge a book by its cover and that goes for videos, too! The colorful cardboard box in which this tape is contained misleads the customer into expecting an animated adventure when, in fact, it contains no drawings or animation whatsoever. A picture or word description of the style of presentation could easily have been included on the package but was not. While the message itself is good, this episode of the series is a low-budget, live action drama (acted out by a cast of children). It opens with Zacchaeus unmercifully collecting exorbitant taxes. He is hated by those whom he defrauds. Progresses to scene at Zacchaeus' home where his children wish that their father would find a more respectable job so that they would have more friends. Ends by showing a repentant Zacchaeus who meets and seeks to please Jesus by vowing to repay everyone from whom he has stolen four times the amount that he stole. This video's low purchase price is attractive, but its low technical quality is not.   ©1988

TREASURES OF THE SNOW

Content: ★★★★★
Technical Quality: ★★★★★

PRIMARY AUDIENCE: Entire Family
STYLE: Drama

LENGTH: 1 hr., 48 min.
DISTRIBUTOR: Children's Media
Productions
818/797-5462

You won't find a better Christian family film than this one! Shot on location in the Swiss Alps, TREASURES OF THE SNOW is a dynamic production that shows how harboring an unforgiving spirit only ends up bringing pain to yourself and the ones you love. Granting forgiveness, on the other hand, brings emotional and spiritual healing. Based on the powerful book by Patricia M. St. John, young Lucien's careless act of childish teasing turns into tragedy when the boy he is teasing falls over a steep mountain cliff and seriously injures his leg. The tragedy leads to bitterness as Annette, the sister of the child whose leg Lucien has damaged, refuses to accept Lucien's attempts to win forgiveness. Eventually, Annette's bitterness drives her to brokenness and, finally, to a willingness to forgive. Other sub-plots are equally compelling. If at all possible, get your church to have a Sunday evening showing of this one in its 16mm format. Then, buy the video and invite non-Christian neighbors over to see it with you. WARNING! You'll need a box of tissues handy during this one. *(Also available on 16mm film.)*   ©1983

WENDY AND THE WHINE

Content: ★★★★
Technical Quality: ★★★★

PRIMARY AUDIENCE: Ages 3 - 8
STYLE: Animation

LENGTH: 27 min.
DISTRIBUTOR: Family Films
314/664-7000

This humorous children's musical cartoon effectively teaches that whining is very much disliked by all who come in contact with the whiner. In WENDY AND THE WHINE, a "whine monster" escapes from little Wendy's mouth whenever she whines too much. She finds that people don't like being with a whiner and eventually even decides that she doesn't like whining herself. However, she doesn't know how to control her bad habit. She tries conquering it by whispering, not talking, and then by taping her mouth completely shut. Finally, she discovers that she needs to ask God to help her conquer her bad habit. Wendy climbs up on her grandmother's lap and together they pray for the Lord's help. Grandma has a good idea—"Think, 'A whine is a loathsome and troublesome thing' before you speak." Concludes with a happy family around the dinner table. This entertaining tape would make a welcome gift for almost every family that has a child in the age category listed!   ©1987

WHAT WIVES WISH THEIR HUSBANDS KNEW ABOUT WOMEN

Content: ★★★★★
Technical Quality: ★★

PRIMARY AUDIENCE: College - Adult
STYLE: Lecture

LENGTH: 1 hr., 22 min.
DISTRIBUTOR: Word, Inc.
214/556-1900

This video should be viewed repeatedly by every married couple and soon-to-be-married single. While the style is very simple (basically consists of Dr. James Dobson speaking before a live audience in an auditorium), viewers are not bothered by the producer's lack of creativity. Dr. Dobson is a masterful speaker who uses humor and emotion-filled real-life stories to help communicate his five-star message. Deals primarily with the subject of depression in women, the ten main causes for it, and how men and women can work to avoid or cure it. Also reveals some of the primary emotional differences between men and women. Buy (don't just rent) this tape, so that you can have it on hand to help minister to others including neighbors, relatives, and church friends. Note: Clothing styles have changed since this tape was produced, but the emotional needs of women are basically the same today as they were then. *(Also available on 16mm film.)*
©1979

WHO DO YOU LISTEN TO?
Sex in the Age of AIDS

Content: ★★★★
Technical Quality: ★★★★★

PRIMARY AUDIENCE: Jr. High - College
STYLE: Documentary

LENGTH: 37 min.
DISTRIBUTOR: Gospel Films
616/773-3361

This powerful "film essay" uses well-acted dramatic vignettes to steer young people toward dating relationships that are free from the trappings of premarital sex. Fast paced and moved along by the thought provoking statements of a young, on-camera host, and by letters sent to and read by Josh McDowell, WHO DO YOU LISTEN TO? is very popular among high school audiences. Includes upbeat songs by the contemporary Christian rock group Petra and the rap group PID (Preachers in Disguise). Because it was produced primarily for presentation in public school settings, it contains no blatantly biblical directives. Instead, it discourages premarital sex by emphasizing the possibility of catching sexually transmitted diseases, especially AIDS. Dramatic sequences effectively encourage youths to wait for that one special person to whom they actually pledge their lives through a marriage ceremony. An excellent discussion starter. *(Also available on 16mm film.)*
©1989

WINDSONGS

Content: ★★★
Technical Quality: ★★1/2

PRIMARY AUDIENCE: Adults
STYLE: Praise & Worship

LENGTH: 30 min.
DISTRIBUTOR: Tyndale Christian Video
708/668-8300

This tape combines video footage of nature scenes with instrumental medleys of 28 different hymns such as "Fairest Lord Jesus," "Amazing Grace," etc. Instrumentation is good, but usually not more than one or two different instruments at a time (not a rich, orchestral sound). The video footage, from the archives of the Christian Broadcasting Network, includes streams, lakes, blowing wheat, whales, snow, and sunsets. Various styles of calligraphy display some of the words to some of the songs. Instrumentation only, no narrator or singer. (This seems to be a low-budget attempt to "duplicate" the high quality Moody/Maranatha! videos THIS IS THE DAY, PERFECT PEACE, etc. Good try, but not in the same league.)   ©1989

WITNESSES OF JEHOVAH

Content: ★★★★
Technical Quality: ★★★

PRIMARY AUDIENCE: Jr. High - Adult
STYLE: Documentary

LENGTH: 58 min.
DISTRIBUTOR: Bridgestone Video
619/431-9888

This documentary is essential viewing for anyone currently in or considering association with the Watchtower Society (Jehovah's Witness). In an investigative style, the producer traces the history of the organization from its first days, penetrating the schemes, scams and false prophecies used in this pseudo-Christian cult to maintain control over its 3.4 million members, and to extend its worldwide empire. A valuable addition to any church video library. Churches should make this one available to their members to view in their own homes as they prepare for these inevitable door-to-door visitors. *(Also available on 16mm film.)*   ©1986

THE WORLD THAT PERISHED

Content: ★★★★★
Technical Quality: ★★★★

PRIMARY AUDIENCE: Jr. High - Adult
STYLE: Documentary

LENGTH: 33 min.
DISTRIBUTOR: Films For Christ
602/894-1300

This award-winning documentary combines real-life cinematography, animation, and special effects. Presents a well-researched concept of what the geography and climate of the earth was probably like prior to the worldwide flood-judgment of Noah's day. The producer's use of a wide range of visuals adds entertainment value to the film which includes a reenactment of Noah's family building the ark, the ark on the stormy sea, the 35,000 animals housed in their cages and stalls on board the ark, erupting volcanoes, modern-day search teams looking for remains of Noah's Ark on Mt. Ararat, etc. It documents flood legends from tribes all around the world which are very similar to the actual scriptural record found in Genesis. This near timeless film is an excellent apologetic for the accuracy and trustworthiness of the first eleven chapters of Genesis. Viewers will marvel at many of the faith-strengthening evidences presented. Excellent evangelistic conclusion. A staple film for church and Christian school video libraries. (*Also available on 16mm film.*) ©1977

YOUR CRISIS PREGNANCY

Content: ★★★★
Technical Quality: ★★★1/2

PRIMARY AUDIENCE: Anyone facing a crisis pregnancy
STYLE: Documentary

LENGTH: 25 min.
DISTRIBUTOR: American Portrait Films
714/535-2189

YOUR CRISIS PREGNANCY was created specifically to help women facing a crisis (unplanned or unwanted) pregnancy. This film sensitively reveals the various stages of prenatal life, summarizes what actually takes place in an abortion procedure, and exposes the potential of post-abortion syndrome and possible physical complications. Contains actual testimonies of four young ladies who have faced the crisis of unwanted pregnancy. Describes some of the avenues of practical help now available (adoption services, etc.). YOUR CRISIS PREGNANCY should be included in every church video library and shown in all high school through adult Sunday School classes to make its content and availability known to the congregation. Such availability will encourage average Christians to use the tape to reach out to those whom they encounter facing a crisis pregnancy. (Suggestion: Type the names, phone numbers, and addresses of area crisis pregnancy centers on a small piece of paper and tape this list inside the video case.) ©1987

VIDEO INDEX

★★★★★ Excellent!   ★★★★ Very Good   ★★★ Good   ★★ Fair   ★ Poor

✳ = Review included in the Videoguide section

TITLE	-- QUALITY RATING --	
	Content	Technical
✳ ACTS: Volume 1	★★★★1/2	★★
✳ A.D. (Abridged Edition)	★★	★★★★★
✳ ADVENTURES OF CHARLIE WANDERMOUSE: Tape 1	★★1/2	★★★
ADVENTURES OF CHARLIE WANDERMOUSE: Tape 2	★★1/2	★★★
✳ AIDS: A Christian Perspective	★★★	★★
✳ AMAZING BOOK, The	★★★★★	★★★★★
✳ AMY GRANT: "Find A Way"	★★	★★★★★
ANGEL ALLEY	★★★★	★★★★
✳ ANIMATED STORIES FROM THE NEW TESTAMENT: Volume 1, "The King is Born"	★★1/2	★★★★★
✳ ANIMATED STORIES FROM THE NEW TESTAMENT: Volume 2, "He is Risen"	★★★	★★★★★
✳ ANIMATED STORIES FROM THE NEW TESTAMENT: Volume 3, "The Prodigal Son"	★★★★	★★★★★
✳ ANIMATED STORIES FROM THE NEW TESTAMENT: Volume 4, "The Good Samaritan"	★★★1/2	★★★★★
Anthony Paul's - OBEDIENCE AND SELF CONTROL	★★★★1/2	★★1/2
✳ Anthony Paul's - SHARING AND KINDNESS	★★★★★	★★1/2
✳ BASICS OF BASS FISHING	★★★★	★★★★
✳ BEAR HUGS: Volume 2	★★★	★★★
BEGINNING GOLF FOR WOMEN	★★★	★★
BENNIE	★★★★	★★★1/2
✳ BEYOND THE NEXT MOUNTAIN	★★★★★	★★★★1/2

TITLE	-- QUALITY RATING --	
	Content	Technical
DIETRICH BONHOEFFER: Memories and Perspectives	★★★1/2	★★★
* BORN AGAIN	★★★★	★★★★★
BY LOVE SET FREE	★★★	★★★1/2
* CARMAN: Radically Saved	★★★	★★★★
* CAUGHT	★★★★1/2	★★★★★
* CHILDREN'S VIDEO BIBLE: "The Beginning"	★★★	★★1/2
A CHRISTMAS CAROL	★★★	★
* CHRISTMAS IS	★★★	★★★★
* CIRCLE SQUARE: Volume 12 "Broken Mirrors"	★★★	★★
* CITY OF THE BEES	★★★★★	★★★★
* CITY THAT FORGOT ABOUT CHRISTMAS, The	★★★	★★★★
CLEBE McCLARY: Portrait of an American Hero	★★★	★★★★
CLOWN-FACED CARPENTER, The	★★★★★	★★★★
* COACH	★★★★	★★★1/2
JAN AMOS COMENIUS	★★★	★★★★
* CROSS AND THE SWITCHBLADE, The	★★★★	★★★★★
* CRY FOR FREEDOM, A	★★★★★	★★★★1/2
* CRY FROM THE MOUNTAIN	★★★★1/2	★★★★★
* C.S. LEWIS: "Through the Shadowlands"	★★★★	★★★★★
DANGER ON THE PIONEER EXPRESS	★★★★	★★★1/2
* DANGEROUS JOURNEY	★★★★★	★★★★
* DAVEY AND GOLIATH: Volume 1	★★★	★★★1/2
DAVEY AND GOLIATH: "Christmas Lost and Found"	★★★	★★★★
* DAVEY AND GOLIATH: "Halloween Who-Dun-It"	★★★1/2	★★★★

TITLE	-- QUALITY RATING --	
	Content	Technical
* DAVEY AND GOLIATH: "Happy Easter"	★★★	★★★★
* DESTRUCTIVE DAVID	★★★★	★★★★
* DISTANT THUNDER, A	★★★★	★★★1/2
* DISTINCTIVELY HUMAN	★★★★★	★★★★★
* DOVETALE'S: "Noah's Ark"	★★★	★★
* ECLIPSE OF REASON	★★★★★	★★★★
* EVOLUTION CONSPIRACY, THE	★★★★	★★★★
* FIRST FRUITS	★★★★★	★★★★
FISHERMAN, The (VCR Bible Game)	★★★	★★
* FLYING HOUSE, The, Volume 4	★	★★★1/2
* GERBERT Series: Volume 5—"Safe in His Arms"	★★★★	★★★★
* GOD OF CREATION	★★★★	★★★1/2
* GOD'S OUTLAW: The Story of William Tyndale	★★★★1/2	★★★★
* GOLF FOR KIDS OF ALL AGES	★★★★	★★1/2
* GOSPEL BILL: "There is No Such Thing As Monsters"	★★★★	★★★
* GREAT DINOSAUR MYSTERY, The	★★★★★	★★★
GREATEST STORY OF ALL, The	★★	★★
GREATEST TALES FROM THE OLD TESTAMENT: "Three Great Leaders"	★★	★★★
* Hanna-Barbera's - CREATION, The	★★	★★★★★
* Hanna-Barbera's - DANIEL AND THE LION'S DEN	★1/2	★★★★★
* Hanna-Barbera's - DAVID AND GOLIATH	★★★	★★★★★
* Hanna-Barbera's - JOSHUA AND THE BATTLE OF JERICHO	★1/2	★★★★★
* Hanna-Barbera's - NOAH'S ARK	★★	★★★★★
* Hanna-Barbera's - SAMSON AND DELILAH	★★1/2	★★★★★

TITLE	QUALITY RATING	
	Content	Technical
HEALING, The	★★★★	★★★
* HEAVEN'S HEROES	★★★★	★★★1/2
* HIDING PLACE, The	★★★★★	★★★★★
* HOOMANIA	★★★★	★★★★
* HOW SHOULD WE THEN LIVE	★★★★★	★★★★
* HOW TO GET BETTER GRADES	★★★★1/2	★★★
* HOW TO KNOW YOU'RE IN LOVE	★★★★	★★
* HUMAN RACE CLUB, Volume 1: "Self Esteem"	★★1/2	★★★★
* HUMPTY	★★★	★★★★1/2
* IMAGE OF THE BEAST	★★★★	★★★1/2
* ISRAEL - GOD'S CHOSEN LAND	★★★	★★1/2
* JESUS	★★★★★	★★★★★
* JESUS' BICYCLE	★★★★★	★★★
* JESUS' LIFE, Tape 1: "His Birth"	★★★★★	★★★
* JESUS' LIFE, Tape 2: "His Ministry"	★★★★★	★★★
JESUS' LIFE, Tape 7: "His Last Week"	★★★★	★★★
* JIMMY AND THE WHITE LIE	★★★★	★★★★
* JOHN HUS	★★★★★	★★★★
* JOHN WYCLIFFE: The Morning Star	★★★★	★★★★
* JONI	★★★★1/2	★★★★★
* JOURNEY OF LIFE	★★★★1/2	★★★★★
JOY OF BACH, The	★★★	★★★1/2
* Ken Anderson's - ELIJAH AND THE FIRE FROM HEAVEN	★★★★	★★★
* Ken Anderson's - JONAH AND THE BIG STORM	★★★★★	★★1/2
* Ken Anderson's - JOSEPH'S DREAM	★★★1/2	★1/2

TITLE	QUALITY RATING	
	Content	Technical
* Ken Anderson's - JOSHUA AND THE PROMISED LAND	★★★★★	★★1/2
* Ken Anderson's - MARY AND JOSEPH	★★★★1/2	★★1/2
Ken Anderson's - MIRIAM AND BABY MOSES	★★★★★	★★1/2
* Ken Anderson's - NOAH'S BIG ADVENTURE	★★★★1/2	★★1/2
* Ken Anderson's - PAUL'S ADVENTURES	★★★★★	★★1/2
* KIDS PRAISE! 4: Singsational Servants	★★★★	★★★1/2
KIDS PRAISE! 5	★★★	★★★
* KINGDOM CHUMS, The: "Little David's Adventure"	★	★★★★★
* LAUREL WITH A MORAL	★★★★★	★★★1/2
* LEARNING ABOUT SEX (Ages 6 - 8): "Where Do Babies Come From?"	★★★★	★★
* LEARNING ABOUT SEX (Ages 8 - 11): "How You Are Changing"	★★★★	★★
* LIFE FLIGHT	★★	★★★
* LION, THE WITCH, AND THE WARDROBE, The	★★★★★	★★★★1/2
* LITTLE TROLL PRINCE, The	★★	★★★★★
* LITTLE VISITS WITH GOD: VOLUME 1	★★★★★	★★★1/2
LOST GOLD MINE	★★★1/2	★★★
* MAKE YOUR OWN MOVIE KIT—Stories From the Bible	★★★	★★
* MARTIN LUTHER	★★★★1/2	★★★★
* MARTIN LUTHER - HERETIC	★★★★★	★★★★★
MASSACRE OF INNOCENCE	★★★★	★★
MATTER OF CHOICE, A	★★★	★★★
* McGEE AND ME!, Episode 1: "The Big Lie"	★★★★★	★★★★★
* McGEE AND ME!, Episode 2: "A Star in the Breaking"	★★★★	★★★★★

TITLE	-- QUALITY RATING --	
	Content	Technical
* McGEE AND ME!, Episode 3: "The Not-So-Great Escape"	★★★★★	★★★★★
* McGEE AND ME!, Episode 4: "Skate Expectations"	★★★★★	★★★★★
* McGEE AND ME!, Episode 5: "Twister and Shout"	★★★★★	★★★★★
* MIKE WARNKE—Do You Hear Me?!	★★★1/2	★★★★
* MORMON DILEMMA, The	★★★★	★★
* MUSIC BOX	★★★	★★★★
NEW MEDIA BIBLE, Part 1: "In the Beginning"	★★	★★★
NIKOLAI	★★★★	★★★★
* ORDINARY GUY	★★★★1/2	★★★★
* ORIGINS - How the World Came to Be	★★★★★	★★★★★
* OUR DWELLING PLACE Series— "The Trials of Jesus"	★★★★	★★1/2
* PARABLES, The: Volume 2	★★★★★	★★★
* PARABLES FROM NATURE: Series 5	★★★★★	★★★1/2
PARADISE TRAIL, The	★★★★	★★★
PEACE CHILD	★★★★	★★★1/2
* PEDRO AND THE BRIGHT CITY	★★★★	★★
* PERFECT PEACE	★★★★★	★★★★★
PETER AND THE MAGIC SEEDS	★	★★★★
* PILGRIM'S PROGRESS (Baptista)	★★★★	★★★
PILGRIM'S PROGRESS (Anderson)	★★★★	★★
* PRODIGAL, The	★★★★1/2	★★★★★
* PRODIGAL PLANET, The	★★★1/2	★★★★
PROSECUTOR, The	★★1/2	★★★

TITLE	-- QUALITY RATING --	
	Content	Technical
* QUESTIONS PARENTS ASK - Dr. James Dobson	★★★★	★★
* QUIGLEY'S VILLAGE: "Be Kind to One Another"	★★★★★	★★★★
* QUIGLEY'S VILLAGE: "Fun Aerobics for Kids!"	★★★★	★★★★
* ROLE OF PRAYER IN SPIRITUAL AWAKENING, The	★★★★★	★1/2
* SAMMY	★★★★	★★★
* SANDI PATTI: "Make His Praise Glorious"	★★★★★	★★★★
* SATANISM UNMASKED	★★★★	★★1/2
SHEPHERD, The	★★★1/2	★★★1/2
* SHIOKARI PASS	★★★★★	★★★
SO MANY VOICES	★★★	★★
* SPACE SHUTTLE JOURNEY	★★★★1/2	★★★★
* SPORTS ODYSSEY, A	★★★	★★★★
* STORMIE OMARTIAN'S First Step Workout Video	★★★★★	★★★★
* STORMIE OMARTIAN'S Low Impact Aerobic Workout	★★★★★	★★★★1/2
SUNDAY SING-ALONG VIDEO	★★★★	★★1/2
* SUNSHINE FACTORY: "Anger"	★★★	★★★1/2
SUNSHINE FACTORY: "Honesty"	★★★★	★★1/2
* SUPERBOOK, Volume 5	★1/2	★★★1/2
* SUPERCHRISTIAN	★★★★1/2	★★★
SURVIVAL	★★★	★★★
* SWISS FAMILY ROBINSON, Volume 1: "The Terrible Typhoon"	★★★★	★★★1/2
* SWISS FAMILY ROBINSON: Volume 3 "Mysterious Island"	★★★★	★★★1/2
* THIEF IN THE NIGHT, A	★★★★	★★★1/2
* THIS IS THE DAY	★★★★★	★★★★★

TITLE	-- QUALITY RATING --	
	Content	Technical
* TREASURE CHEST Series: "Zacchaeus - Little Man up a Tree"	★★1⁄2	★
* TREASURES OF THE SNOW	★★★★★	★★★★★
* WENDY AND THE WHINE	★★★★	★★★★
* WHAT WIVES WISH THEIR HUSBANDS KNEW ABOUT WOMEN	★★★★★	★★
* WHO DO YOU LISTEN TO?—Sex in the Age of AIDS	★★★★	★★★★★
* WINDSONGS	★★★	★★1⁄2
* WITNESSES OF JEHOVAH	★★★★	★★★
* THE WORLD THAT PERISHED	★★★★★	★★★★
* YOUR CRISIS PREGNANCY	★★★★	★★★1⁄2